The Conservation of Butt
past and present

by

John Feltwell

Wildlife Matters, Battle, England

Published by Wildlife Matters, Marlham, Henley's Down, Battle, East Sussex, TN33 9BN, UK

Some other titles by John Feltwell

Animals and Where They Live (Dorling Kindersley)
Bugs, Beetles and Other Insects (OUP)
Butterflies and Moths (Dorling Kindersley)
Butterflies and Other Insects of Britain (Reader's Digest)
Butterflies of Europe (Dragon's World)
Butterflies of North America (Dragon's World)
Countryside Conservation (Ebury Press)
Discovering Doorstep Wildlife (Hamlyn)
Encyclopaedia of Butterflies of the World (Quarto)
Meadows - A History and Natural History (Alan Sutton)
Studying Butterflies (Learning Through Landscapes)
The Biology of the Large White Butterfly (Dr. W. Junk)
The Natural History of Butterflies (Croom Helm)
The Story of Silk (Alan Sutton)

First edition 1995   ISBN 0 907970 02 8
© 1995 John Feltwell / Wildlife Matters (text)
© 1995 Brian Hargreaves (illustrations)
Design, layout and desktop published by Wildlife Matters
Printed in Great Britain by The Chameleon Press, 5-25 Burr Road, London, SW18 4SG on paper from sustainable forests.

This publication may not be reproduced, stored or transmitted electronically, in any form or by any means, without the prior permission in writing of the publishers. This does not affect the provisions laid out in the UK Copyright Designs and Patents Act, 1988 regarding fair dealing for the purposes of research or private study, or criticism or review. The publisher has made every effort to ascertain the accuracy of the information contained in this book, including consultation and liaison with appropriate bodies, and cannot accept any legal responsibility or liability for any unwitting errors, omissions, or implications, implied or otherwise, notwithstanding any unforeseen liability arising through *force majeure*, that may have been made, or may arise in the future, by any means. In a work of this magnitude (95,000 words) it is inevitable that mistakes will have crept in. John Feltwell welcomes comments, additions, amendments and updates, with supporting documentation, at the Wildlife Matters address.

A catalogue record for this book is available from the British Library and is included in the Whittaker's List.

In memory of my parents Edna and Ray

Illustrations by

Brian Hargreaves of Rye, East Sussex

some other titles illustrated by Brian Hargreaves

Buczacki & Harris's Pests, Diseases & Disorders of Garden Plants
Carter's Caterpillars of Butterflies & Moths
Chinery's Butterflies and Day Flying Moths
Chinery's Gem's guide to Butterflies
Dulcie Gray's Butterflies on my mind
Feltwell's Butterflies of Britain & Europe
Feltwell's Butterflies North America
Harley Books, Moths and Butterflies of Britain & Ireland, Vols 1,2,9,10
Higgins & Riley's Butterflies of Europe
Riley's Butterflies of the West Indies
and, Butterfly postage stamps for nine countries

List of illustrations

1	Ploughing ancient meadows	1
2	Natural History Museum, London	11
3	Butterfly biodiversity	27
4	Chequered Skipper	39
5	English Swallowtail in the fens	45
6	Large copper larvae surviving immersion	55
7	Large Blue habitat, early twentieth century	71
8	D.F.B. Stucley's reserve sign	74
9	Map of North Cornwall and Devon coast	77
10	Flora of Large Blue habitats	89
11	Heath Fritillary habitat in coppice woodland	97
12	The Aurelians (1766) print	110
13	The New Aurelian	111
14	Development versus Butterflies	113
15	Lulworth Skippers at Lulworth Cove	129
16	The collectors' classic kit	153
17	The Duke of Burgundy	163

Contents

List of illustrations   *iv*
Acknowledgements   *vi*
Preface   *vii*
Standards   *ix*

CHAPTER 1	Introduction	1
CHAPTER 2	Historical Background	11
CHAPTER 3	Why Conserve Butterflies?	27
CHAPTER 4	Flagship Species	39
CHAPTER 5	Conservation of the Swallowtail	45
CHAPTER 6	Conservation of the Large Copper	55
CHAPTER 7	Conservation of the Large Blue	71
CHAPTER 8	Habitat Management	97
CHAPTER 9	Threats to Butterflies	113
CHAPTER 10	Butterflies and the Law	129
CHAPTER 11	Voluntary Codes and Practices	153
CHAPTER 12	Nature Reserves	163

Appendix 1	Species action plan for the Large Blue Butterfly	173
Appendix 2	A code for insect collecting 1972	175
Appendix 3	Summary of RSNC policy on introductions	178
Appendix 4	A code for dealers	179
Appendix 5	Insect re-establishment - a code of conservation practice	181
Appendix 6	Butterfly Conservation's code on butterfly releases	188
Appendix 7	Butterfly Conservation's national conservation strategy	190
Appendix 8	Nature reserves for butterflies	192
Appendix 9	The British species	197

Acronyms   199
Glossary   201
Bibliography   203
Index   229

Acknowledgements

The following have been invaluable in my quest for information, either sorting out queries, offering opinions or providing detailed local information: Professor R. J. Berry of University College London, Mr. B. O. C. Gardiner of Cambridge, Professor J. N. R. Jeffers formerly of Merlewood Research Station, Graham Howarth formerly of the Entomology Department at the Natural History Museum in London, Professor M. G. M. Morris, Honorary President of the Joint Committee for the Conservation of British Invertebrates, I. K. Morgan from the Countryside Council for Wales, Dr. John Muggleton of Slough, Dr. T. A. Rowell of Dyfed, Mrs Jacqueline Ruffle, Librarian at the Royal Entomological Society of London, Adrian Spalding, Director of the Cornish Biological Records Unit, Dr. Colin Studholme of the Gloucester Wildlife Trust, and Ken Willmott of Merton Park. There are many others with whom I have discussed various ideas and to whom I am also very grateful. At Butterfly Conservation, Dr. Martin Warren, Conservation Officer, Paul Kirkland, Assistant Conservation Officer, Gary Roberts and Patrick Roper both of Archmain Ltd. and both Press Officers for Butterfly Conservation, have provided lots of background information on that burgeoning society; at English Nature (Peterborough), Dr. A. Deadman and Dr. Martin Drake have provided statistical information or comments, and Dr. David Sheppard has provided a continuous stream of background information and forthright comment. Miriam Rothschild has remained a stalwart of information on the formative years of conservation earlier this century and has provided information about her conservation-minded father and uncle. I would particularly like to thank David Carter in charge of British lepidoptera at the Natural History Museum who patiently read the entire manuscript and made invaluable recommendations from which the text benefited. Brian Hargreaves, the illustrator, patiently followed my ideas through to reality, and I am very grateful to him for producing such a fine set of line drawings. John McNaughton provided invaluable technical assistance during production. Finally, I would like to thank my wife for her understanding whilst this book was being realized.

Preface

This book is devoted to the butterflies of Britain. There is perhaps more emphasis on butterflies in England than elsewhere in Britain, since this is where there is the greatest richness of butterflies, and where most of the work has been done. It has been tempting to include details about the conservation of butterflies on the Continent, but with nearly 600 references to the conservation of butterflies in Britain, only those few with direct relevance are included.

Britain is unique in Western Europe in having such a strong Atlantic influence which tempers any extremes in climate, and produces the fascinating ecological relationships of butterflies. Continentals look upon Britain as a sort of off-shore field station where the ecology of butterflies is sometimes atypical, and where butterflies are problematic to conserve. The evolution of butterflies in Britain has taken its own course slowly; after all there are no endemic species, just a few endemic subspecies. There is still a European influence not to be ignored, dating back to the last Ice Age when there was a land connection between Britain and the rest of Europe; some British populations are still restocked annually through immigration. Some of Britain's butterflies are at the edge of their range in Western Europe, and it is noteworthy that re-establishment of species from the Continent has been used to augment Britain's dwindling populations.

The phenomenal growth in enthusiasm for butterflies as seen in the soaring membership of the Butterfly Conservation society, now with over 10,000 members, attests to local interest in butterflies, and explains a lot of the current enthusiasm and advances in butterfly conservation. It is as if, with a declining national list, the enthusiasm for butterflies increases proportionally. This book is aimed at all those who are specifically interested in butterflies, whether at an amateur or professional level.

Britain leads the world in butterfly conservation, and other countries use the knowledge gleaned in our woods and meadows, heaths and downs, to conserve their own species. Most prominent amongst the flagship species has been the decades of work carried out, first on the Large Copper, then the Large Blue and Swallowtail.

Lessons learnt on the English Large Blue are now used as a template for conservation on the Continent. Whilst Britain's flagship species have traditionally had a bigger share of resources, the majority of British species have had no conservation management applied to them, and they continue to struggle for survival in a world of declining habitats.

This book is about all those butterflies that have had some sort of conservation applied to them; it is not a 'how-to' book or management handbook, nor is it a book about the ecology of butterflies. Understanding their ecology is not the same as understanding how they should be conserved. An overview of the natural history of Britain's butterflies was dealt with in one of my previous books, *The Natural History of Butterflies*, and this book is complimentary to it.

The need for an overview on the state of Britain's conserved butterflies is well overdue in the light of the massive declines experienced by so many species. Alarming decreases in numbers of butterfly species was highlighted by Butterfly Conservation in 1995 when they declared 25 species as threatened. Species have 'withdrawn' from both eastern England and southeast England where habitat destruction has played a major role. The word 'withdrawing' - much used in the study of butterflies - is simply a misnomer for saying that they have become extinct locally.

An enormous amount of research has been put into unravelling this history of the conservation of butterflies in Britain and in presenting an overview of the situation. To this end I have been aided by fine assistance from various quarters but responsibility for any inaccuracy rests entirely with me. I trust that you, the reader, will find plenty of avenues of interest; indeed, it is the first time that the collective information on, for instance, the conservation of the Large Blue has ever been brought together in one place. With so much ecological information still lacking on most of Britain's butterflies, and knowledge of the flagship species still wanting, the conservation of Britain's butterflies will have to be modified in the light of future advances. In Britain where the bulldozer has caused so much habitat loss, it is, paradoxically, the best conservation tool in managing some butterfly habitats. There will still remain intrigue surrounding some of Britain's rarest butterflies, and the identity of Mr X, and of Site X or Y or M will perhaps be clarified in the future.

John Feltwell
Sussex and the Cévennes, May 1995

Standards

All butterflies mentioned conform to J. D. Bradley and D. S. Fletcher's (1979) *A Recorder's Log Book of British Butterflies and Moths* (Curwen Press). Names and dates of deceased entomologists follow Pamela Gilbert's (1977) *Bibliography of Deceased Entomologists* (British Museum of Natural History, London). All plants mentioned conform to J. G. Dony, S. L. Jury & F. Perring (1986) *English Names of Wild Flowers* (Botanical Society of the British Isles). The term 'Britain' is used to denote England, Wales, Scotland and Ireland, and the term 'United Kingdom' is used in the context of the legal framework which includes England, Wales, Scotland and Northern Ireland. 'Continental Europe' is used to describe that part of Europe other than Britain.

Permissions

All the major conservation bodies concerned with butterfly conservation were contacted during the course of the preparation of this book, as well as all the Wildlife Trusts. Special thanks go to English Nature at Peterborough for permitting me to consult their confidential reports on butterflies from which details appear in this book; this was kindly negotiated by David Sheppard. The following gave permission to use their material: the Amateur Entomologists' Society for permission to reproduce their Dealers' code, Butterfly Conservation to quote freely from their Press Releases and their Codes, The Joint Committee for the Conservation of British Invertebrates to consult and quote from their Minute books and documents lodged at the Royal Entomological Society of London, The Wildlife Trusts for their SPNR code, and the Royal Society for the Protection of Birds for permission to quote the Action Pan for the Large Blue. The Librarians at the Royal Entomological Society of London and the Linnean Society, Mrs Jacqueline Ruffle and Gina Douglas respectively gave invaluable support.

CHAPTER ONE
Introduction

'Conservation biologists are prone to preservationism.' Robert Berry, 1992

Butterflies are to entomologists, what birds are to general natural historians. They are eagerly applauded as symbols of a healthy environment, are extraordinarily beautiful, and are relatively easy to study. Butterflies have panache and charisma and are the perfect embodiment of the natural world in all their splendour. To put them in perspective, however, butterflies represent only a quarter of one percent of the whole British insect fauna (some 22,500 species, Shirt, 1987) **(1)** but they get more than their share of acclaim, deservedly.

That butterflies have stolen the limelight is relatively unfair on their insect partners in the living world, moths - all 2,340 species of them in Britain. That makes British butterflies species outnumbered 40 to 1 by moths. Some moths, like the burnets, are just as colourful as the butterflies, the Emerald Moth just as delicate. Although there is no moth society to cater exclusively for their

Introduction

conservation interests, a significant proportion of the swelling masses of Butterfly Conservation (BC) membership are also keen moth enthusiasts. In the context of the subject of this book, the conservation of butterflies overlaps a little with the conservation of moths, since in some habitats, generally speaking what is good for butterfly conservation is good for moth conservation.

Britain has such a tiny selection of butterflies, a mere 59 breeding species, that one wonders whether butterfly enthusiasm is inversely proportional to butterfly paucity. Just as a tropical country may boast about its colourful butterflies of the rainforest, Britain is justly proud of its compact and credible species list which can actually stretch to 70 species including migrants. The 59 breeding species are all that Britain has now, but each one is still worth conserving.

There was a time though when other species more familiar on the Continent were on the British list of butterflies. One such species was the Scarce Swallowtail which was regarded in the eighteenth century as being native to Britain, but has since dwindled to extinction. (2) There has been a decline in Britain's butterflies since the end of last century, culminating in two substantial losses in the 1970s. England lost the Chequered Skipper in the mid-1970s, although it fortunately survived in Scotland. England then lost the Large Blue in 1979.

While Scotland has no other butterfly species exclusive to itself (subspecies excluded), it can boast 36 species (Thomson, 1980), three of which have larger areas of distribution in Scotland than elsewhere (Northern Brown Argus, Scotch Argus, Mountain Ringlet). According to Thomson's calculations, Scotland has had a recorded total of 48 species (excluding 20 suspect species) since 1800, which is not far short of the British figure for the same period. Almost a quarter of the resident species of Scottish butterflies had been lost up to 1980, most of these in the east of the country (i.e. those species that are less tolerant to oceanic climates). Curiously the shrinkage of English populations of many butterflies to strongholds in the west is also now apparent, especially in the case of the larger fritillaries.

The number of butterfly species found on 52 of Britain's islands has been studied by Reed (1982). The Isle of Wight comes out top with 37 species, followed by Anglesey with 29 and Mull 20. In places like the Isle of Man, which has about 17 species, there is a general lack of true woodland species, most being distributed around the coast. Reed's data are very useful as a basis for formulating future strategies for butterfly conservation.

Introduction

CONTINENTAL INFLUENCE

Britain is inescapably part of the European continent, both in terms of ecological and climatic influences, apart from being linked via the shallow continental shelf. One has only to glance at the distribution maps of the British species (e.g. Thomas, 1986) to see how southerly based they are, to understand this affinity to the Continent. About 30 species have a southerly distribution, representing about 50% of the British list. See Dennis (1977) and Thomson (1980) for accounts of butterfly colonisation of the UK and Scotland respectively.

One has therefore to weigh up this Continental influence, with the effects of isolation when considering the conservation of butterflies in Britain. There is no butterfly species in Britain that is not found anywhere else in the world or on the Continent, which is a pity, as this would be a nice national claim. Britain has sub-species endemic to Britain, but no endemic species. The lack of any endemic species in Britain reinforces the notion that Britain is inescapably part of the Continent, and that regional isolation has not been in operation long enough to produce endemic species status. Even the Scotch Argus is found from Scotland through Europe to the Urals and Caucasus mountains. Taking a trans-continental perspective, there are, surprisingly, about 50 species of Europe's 400-odd butterfly fauna that also occur on the North American list, demonstrating both the powers of butterfly dispersal (via the Arctic land bridge) and translocation by people.

The use of English common names of butterflies, instead of Latin ones, is of course of limited use in a world context, but it does serve to focus the mind on particular places and ideal butterfly habitats which is the basis of astute butterfly conservation. There are thoroughly British-sounding butterflies such as the Lulworth Skipper, the Bath White and the Camberwell Beauty, but they are all found on the Continent in larger populations with large distributions, or as in the last species, also in North America, where it loses its British tag and is called the Mourning Cloak.

A good proportion of British butterflies arrive each spring and early summer direct from the Continent. The Painted Lady does not overwinter here, and the Red Admiral, until recent years, did not regularly over-winter in Britain. Now it does, but the populations of both of these species are augmented by, and in some cases originated from, stock flying in from the Continent in the spring and summer. The immigration of Continental specimens is a life-line to 'British species' and any inter-change of genetic material between residents and immigrants may increase vigour and hardiness, or at least widen the gene pool.

Some of the members of resident British populations have their own characteristics for which they have been given subspecies status, and persist as typically British subspecies. Cases in point are the extinct Large Copper (*Lycaena dispar dispar*), and the English Swallowtail (*Papilio machaon britannicus*). It is

Introduction

typical of subspecies that they may exhibit differences in colour, pattern, size of adult, larval foodplant and behaviour, some of which these two English subspecies exhibit, compared with other subspecies on the Continent. Where immigrants breed with residents however one would expect a conformity of colour and patterns. This has happened in the Large White (Feltwell, 1982) but has not widely occurred amongst other British species.

Britain, as an island, encourages the processes of isolation, which means that non-migratory animals, or plants, have only their unreplenished populations within which to breed. This is not always advantageous for the species since populations may dwindle to dangerously low levels. In-breeding can result in the expression of unfavourable genetic traits, and the species may die out due to genetical problems. The lesson for butterfly conservation is always to deal with the largest populations of butterflies in any habitat, and to recognize that small isolated populations may not be viable in the long term.

People enthuse about the British butterflies for many reasons, not least because they are very attractive and are symbols of the countryside. Very few species are pests, such as the Large White or the Small White, yet these are welcomed more in gardens of butterfly-lovers than banished from kitchen gardens. The populations of many of these whites are topped up by immigrants arriving from the Continent each year. Potentially these two species could cause lots of economic damage if they were not kept under control by the widespread use of insecticides. (3)

It is often true that the rarer British butterfly species can be found in larger numbers on the Continent, and some people resident on the Continent use this as an argument to play down the conservation effort on these species in Britain. As explained above the influence from Europe is very real in an evolutionary and historical sense, but there are threats to British butterflies from the Continent which we have to consider alongside our own particular problems, especially habitat loss.

THREATS

European directives are in place to ensure that the wilder parts of Europe (including Britain of course) are covered by expanding transportation infrastructure in the next few years. Such expansion of the road network in Britain is already being felt with the creation of cross-country networks and the linking of by-passes. There are other impacts from air pollution, since this is pan-European in effect and very difficult to control.

Habitat loss is the greatest threat to butterflies in Britain. Poisoning of the environment comes a close second caused by the widespread use of

Introduction

agrochemicals, particularly insecticides which can kill butterflies on contact, whilst fertilisers and herbicides remove their larval foodplants. Although there is no direct evidence that butterflies have suffered population declines since the 1940s because of the use of agrochemicals, this is likely to be the case. The effects of agrochemicals on butterflies are difficult to ascertain because dead butterflies are not conspicuous and easy to collect like a dead otter or osprey.

The virtual impossibility of finding evidence showing environmental effects on butterflies is very worrying in a legal framework, since Community law (which, in issues of conflict, always has supremacy over national law) is now working towards a general acceptance that apparent pollution of the environment is safe provided no-one can prove that it has any adverse effect on species. Or, to put it another way, it is perfectly OK to carry on polluting the environment just so long as no damage to wildlife and habitats can be proven by authorities or individuals.

Whilst threats to butterflies continue, even escalate, there is much concerted effort to conserve Britain's butterflies, mostly carried out by Butterfly Conservation. Although it might appear that conservation of Britain's butterflies is geared towards a few target species, via the official Species Recovery Programme, at grass roots level of conservation in the countryside ordinary people are doing a lot for butterflies in general; in other words there are many endeavours to promote all of the 59 British species of butterfly. Many butterfly nature reserves now boast over 25 species of butterfly.

The key to successful butterfly conservation is getting the habitats and foodplants right. Some butterfly species are easier to conserve since they do not have specialized feeding demands. There are in fact many common species which will always prosper, regardless of what humans do to the environment, since they have very common habitat requirements.

Butterflies belonging to the four big butterfly families are foodplant 'generalists' in a convenient sort of way. The whites exploit the cabbage family (Cruciferae), different groups of nymphalids exploit the nettle family (Urticaceae) or the violet family (Violaceae), the lycaenids exploit leguminous plants (Leguminosae) and the skippers exploit the grass family (Graminae). There is much overlap the more one looks into butterfly-plant relationships, for instance the browns (part of the extended Nymphalidae) have evolved to use various grasses, like many skippers. It is natural that the most numerous butterflies, the whites, browns and skippers are all fairly catholic in their choice of larval food plants. And it is no coincidence that many of the key conservation species are those which have restricted food sources.

But evolution between butterflies and plants is not that simple. Some butterflies have got themselves metaphorically up a very tight evolutionary corner, with absolutely no room for adaptation. In general terms there is little evidence that such butterflies will move onto another unfamiliar food source when the former

Introduction

one disappears. Some butterflies can move onto alternative food sources, but only when their usual one is severely limited.

The Large Blue is an example of a highly specialized species which is living on an evolutionary knife-edge, since it is has both a herbivorous and carnivorous development stage, but this is also the avenue that over 100 species of lycaenids worldwide (to give a wide perspective than just Britain) have taken. Large Blues worldwide, and lycaenids in general, are successful because they have exploited open tracts of land over a wide area. The problem for these lycaenids in Britain is that there is enormous human pressure on the small area of suitable land available. Of the worlds butterflies the lycaenids are the richest group of species, representing some 40% of the total number of butterfly species (Vane-Wright, 1978). Their specialized lifestyle, however (Chapter 7), means that when environmental impacts impinge upon their habitats they have very little room to manoeuvre. Habitat change helped to wipe out the Large Blue in Britain in 1979.

PRESERVATION versus CONSERVATION

Enthusiasm for butterflies lives off itself, and is set to increase. The die establishing the criteria for butterfly conservation has been well cast - a common desire to see lots of butterflies in town and country. Their demise is difficult to follow in such a country as the United Kingdom, which has about 58 million people in a relatively small area (France has about 57 million people, but is three times larger, and the density of people per square kilometre is 104, compared to 236 in the UK (European Community, 1992) and when the demand on land is acute, and is set to become more so.

The enthusiasm for butterflies in Britain is well established, so it is natural that societies have been established to protect and conserve these insects. When we see a suitable habitat burgeoning with butterflies we like to keep it just like that and make it a reserve; but that is when conservation management troubles begin. For when a reserve is established, and a nature reserve sign is driven into the ground, it is in effect a statement saying 'we will have to fight nature'. From that moment on, nature will have to be stopped from what it does best - succession of vegetation - so that the butterflies which like open habitats will prosper. The dilemma for the reserve managers is how to stop the encroachment of nature and keep the butterflies. The brambles and thickets have to be rigorously controlled in order to keep areas specially open for butterflies. That is one way of looking after the countryside, based on restricting nature to meet the needs of one group of animals. Orchid enthusiasts might manage the same reserve in a different way, and ornithologists might make their own stamp on the countryside with their own 'bird landscapes' although they have made recent strides in accommodating

Introduction

butterfly conservation on their reserves. All of these examples serve to show that in Britain we are set in our ways on setting criteria for nature conservation.

There is, however, a completely different way of managing the same piece of land, but one that is hardly used in Britain because of the pressure on land; and that is not being restrictive about those criteria mentioned above. We cannot afford to do this in Britain, since, having discovered a prime habitat for the Duke of Burgundy or for the Heath Fritillary, we would be most distressed to see it revert to some other kind of butterfly-alien habitat within a few years. But this is the kind of criteria set in other countries where land is not such a premium, and you can afford to lose one small area since another will (hopefully) appear somewhere else for the same species. It is worth drawing attention to the scholarly overview of conservation of butterflies in Europe by Otakar Kudrna in 1985 which gives information on the sorry state of the environment in respect of butterflies.

Differing views are seen amongst conservationists regarding the best way to manage habitats for butterflies (Hambler and Speight, 1995). In a thought-provoking article these scientist expose the contentions and logic behind much conservation strategy used traditionally in Britain, and have found it wanting and lacking in supporting evidence. Although considering the wider field of invertebrate conservation, Hambler and Speight recognise that butterflies (as well as flowering plants and birds) have received attention greatly disproportionate to the number of British species present. They question the notion that looking after the butterflies of a habitat will automatically ensure the well-being of numerous other invertebrates. This they find as nonsense, since 'most British invertebrates are small than a butterfly's eye' and live in habitats totally unfrequented by butterflies. They also draw attention to the fact that neglected coppice woodland actually has more ecological niches that a recently coppiced woodland. But opening up of woodland does have its virtues, it depends on what sort of invertebrate you are most interested in.

Having established our specific butterfly criteria, we have to ask ourselves whether we want 'conservation' or 'preservation'. The word *preservation* is a rather abrupt term, used more in North America, for the same general purposes we might use conservation. It implies that nature is going to (have to) stand still because of our management of the habitat. It lacks any context of the future. Preservation is like a hair spray - we want it just like this to hold the habitat in place 'as is' - but conservation is a more malleable concept which implies a more integrated, non- rigid approach. It allows, in a large-sized reserve, for a mosaic of land use, as one might find in a traditionally coppiced wood, or a coppiced wood that is managed for butterflies (e.g. Blean, Canterbury, Kent). The total nature reserve may be a 'preserve' or even a 'monument' as the Americans would say, since it preserves on site a variety of wildlife. Unfortunately a noun for a

Introduction

conservation reserve does not exist in the English language, since a 'conserve' and a 'preserve' have fairly similar sounding culinary connotations. We should stick to having plain old nature reserves which conserve Britain's wildlife.

In essence the best way forward is through 'conservation', not 'preservation' (although the latter has many virtues). If the habitat is properly managed, the flora and fauna is usefully accommodated. 'Conservation' strikes at this basic concept, it is by various definitions: the planned management of natural resources, the retention of natural balance, diversity and evolutionary change in the environment. All this bodes well.

'Preservation' on the other hand has an individualistic approach, a last (management) stand in the face of impending extinction, and it is by definition: the maintenance of individual organisms, populations or species by planned management and breeding programmes'. Considering that some butterfly reserves in Britain have been established around a single rare butterfly species, Britain has, by this definition, already started along the road to preservation, or butterfly 'preserves', although they are not called that yet. Britain's butterfly conservation strategy is therefore particularly sharpened along the path of 'preservation' of what it has left. Conservation is also a major part of that endeavour.

There is no doubt that Britain is a leader in the world of butterfly enthusiasm and conservation. There are more butterfly enthusiasts in this small island than there are, for instance in Latin America (which might have 20 times the number of butterflies). We already know more about (but not enough about) our species than many other countries know about theirs, especially those countries with rainforests. We are in a position to know much more about our own species, but we have let some slip away in recent years through lack of simple observational research (e.g. foodplants or aspects of butterfly ecology or behaviour). We are in the process of a desperate attempt to pick up the ecological minutia about precise habitat requirements so that we can manage them better, but all the time our fragile butterflies are at the mercy of oceanic and continental factors, and of course, humans. Time is running out fast to save our national butterfly heritage. We must not get bogged down in conducting too many ecological investigations, when people in the field are getting on in their own way of conserving butterflies using the best method they have evolved.

The conflict between conservation and preservation is best summed up from one of Professor Sam Berry's more recent, and astute, interpretations of ecological genetics: 'Conservation biology is too full of attempts to rescue species or ecosystems which are transients in the dynamic procession of life, such as the enormous efforts that go into preserving island endemics; conservation biologists are prone to preservationism whereas their science demands a practice of dynamic management which has to accept that extinction is a natural process' (Berry, 1992).

Introduction

Notes

1. It is thought that the true figure of invertebrates in Britain will be around 30,000 excluding truly marine forms (Stubbs, 1983).

2. In southern Europe the Scarce Swallowtail (which belies its name of 'scarce' and is in fact common), graces the countryside from the hottest days of early spring, well through the blistering heat of the summer, and into the autumn as two generations. It needs no special protection since in some areas it is commoner than the Cabbage White, and breeds on Blackthorn (Prunus spinosa), Hawthorn (Crataegus spp.) and fruit trees. In Britain the Scarce Swallowtail could be the subject of a re-establishment scheme (which would no doubt be controversial) if anyone became so inclined. If long hot summers become something of a regular event this Mediterranean beauty, once on the British list, could do well.

3. The Black-veined White was claimed by the noted lepidopterist, L. Hugh Newman to have been 'to some extent an orchard pest' (Newman, undated, p.193) but it died out in about 1914. However, the general feeling now is that the Black-veined White was never a serious pest in Britain. The butterfly remains widespread in different parts of the Continent, but has also become extinct in parts of the north of France, Holland and Belgium (Lhonoré, pers. comm. 1993); in rural southern France populations can be huge (Feltwell, 1983).

CHAPTER TWO
Historical Background to Butterfly Conservation in Britain

'The place had evidently got known and had been thoroughly worked. It was trampled down in all directions, cigarette ends and match ends were scattered about, but the worst feature was that those ant-hills which were visible had been systematically opened up for pupae - the tops had been cut off and were lying alongside.' Frank Labouchere, 1935

ORIGINS OF BUTTERFLY CONSERVATION

An enormous amount of enthusiasm for the conservation of butterflies was promoted by Lord Rothschild and his brother, Nathaniel Charles Rothschild (1), in the early part of the twentieth century. This inordinate enthusiasm for butterflies was unusual then, and has only been re-kindled in recent years by the

Historical background

emergence of Butterfly Conservation. Charles Rothschild established the Society for the Promotion of Nature Reserves (SPNR) which was later incorporated by Royal Charter in 1916 and became the Royal Society for Nature Conservation (RSNC). This has gone on to be one of the largest non-governmental organizations (NGOs) in the country and is now known as 'The Wildlife Trusts' (2). In its early days the SPNR was interested in all sorts of insects, but particular attention was focused on butterflies, and this was often manifest through other groups which the SPNR supported.

So it was that on 14 January 1924, the Central Correlating Committee for the Protection of Nature (CCCPN) was launched under the auspices of the SPNR, whose task was 'correlating the work of the various societies in Great Britain in regard to the protection of animal and plant life, the promotion of nature reserves and kindred objects.' (SPNR, 1925). It soon changed its name to the British Correlating Committee for the Protection of Nature (BCCPN). The initiative for such a correlating committee - a society with the same aims as the present day Wildlife Link - was borne out of the International Congress for the Protection of Nature which had convened in Paris in 1923.

The next stage in the conservation of butterflies was when the British Correlating Committee for the Protection of Nature (BCCPN) was requested by the Council of the Entomological Society of London on 3 June 1931 to set up a special committee to protect British butterflies. This was the start of the very important and influential Committee for the Protection of British Lepidoptera (CPBL) (Entomological Society, Minutes, 3 June 1925). (3)

Whilst all this 'official' conservation was going on, centred around the Royal Entomological Society of London, there was other conservation work being carried out by other individuals. Most notable was the work of Mr L.W. Newman (4) who, according to a recent account of the famous Bexley butterfly farm by Brian Gardiner, was responsible for some early practical nature conservation (Gardiner 1993). An account is given by Gardiner (citing a reference by W. S. Berridge in 1915) of how Newman moved a Marbled White colony being threatened with annihilation from no less than 17 professional collectors, from one meadow to another out of harms way. Although Newman was a highly respected professional breeder of native British species, the butterfly farm concept was one of conservation of species as well as being a commercial venture; however, as Gardiner points out, the very first butterfly farm in Britain pre-dated the founding of Newman's by 10 years when one was established at Scarborough by H. W. Head in 1884. Major clients of Newman's included Walter and Charles Rothschild.

Historical background

CONSERVATION COMMITTEES

Since the turn of the twentieth century butterflies in Britain have been considered by no less than a dozen conservation committees, but only a handful of species have repeatedly received such honoured attention. One has to wonder whether all this concentrated effort on the conservation of the Large Copper, Large Blue and Swallowtail has furthered conservation generally?

The role of the Royal Entomological Society of London (and that of the Entomological Society before it) in the conservation of butterflies has historically gone from one of very keen involvement, to one of dissipated energy. The Entomological Society became involved with butterfly conservation first through the energies of Charles Rothschild up to his death in 1923 (5), and then with Lord Rothschild as part of the Committee for the Protection of British Lepidoptera (CPBL). This was an official organ of the Entomological Society, established by its Council with the Society's rooms made available for hospitality. Thereafter, there was some debate as to the status of this Committee, was it an independent committee of entomologists, or was it a Committee of the Entomological Society? One of its duties to meet its role of acquiring nature reserves was to set up The Nature Reserves Investigation Committee (NRIC) a sub-committee which was in existence by 1936.

The Entomological Society, through the Standing Committee for the Protection of British Lepidoptera, and through the determination of Walter Rothschild, was therefore responsible for conservation measures relating to several species of butterfly and moth. The Entomological Society hosted the first meeting of the Committee for the Protection of British Lepidoptera on 25 September 1925. The formation of this committee by five men in September 1925, is as fundamental to the history of butterfly conservation, as the formation of the National Trust by its team of three in 1895. In the Chair at this first meeting was Lord (Walter) Rothschild with his four members: H. M. Edlesten, J. C. F. Fryer, N. D. Riley and W. G. Sheldon (6). The aim of the committee was 'as far as possible introduce threatened species into new districts, and also create reserves where suitable pieces of land could be obtained for this purpose. The Committee would sit four times a year, twice in the 'collection season' and twice in the autumn. Lists were to be drawn up of species needing protection, and a move was made to publish articles in 'The Times' to gain publicity. The only example of threatened species mentioned at this first meeting were (in the order in which they appear in the Minutes) the Large Blue, Heath Fritillary, Glanville Fritillary, Wood White, Black-veined White, Large Copper and the Mazarine Blue. Clearly the Committee had got their priorities right; one species had already become extinct, the others were variously in trouble, some still are today.

Historical background

As time progressed and other conservation organizations were growing or being established (such as the Commons and Rights of Way Society (CRWS) and the Council for the Protection of Rural England (CPRE), the National Trust (NT) and the International Union for the Conservation of Nature (IUCN)' officials of the Committee for the Protection of British Lepidoptera were dispatched to these other groups to represent the Entomological Society or the CPBL. Gradually the conservation role of the Entomological Society was dissipated in the direction of other specific and general conservation groups. It is interesting that during this time, there was more use of the word 'preserve', as in 'The Society's preserve' and of the word 'preservation' than is used today.

The Committee became increasingly involved with matters other than the conservation of lepidoptera, so it was only natural that the name of the Committee should be changed to reflect this wider appreciation of the entomological world. On 11 June 1931 the CPBL became The Standing Committee for the Protection of British Insects (SCPBI) which thereafter cared for the conservation of all the British insects. Then in 1951 it was reconstituted as The Conservation Committee of the Royal Entomological Society of London (CCRESL) (Southwood 1951).

This Committee worked steadily on until it was absorbed in the Joint Committee for the Conservation of British Insects (JCCBI) which was established in 1968. It was only natural that this de-emphasis from lepidoptera to British insects would occur since, from an early period in its history the Committee for the Protection of British Lepidoptera involved itself with whole habitats and interests in other groups of insects such as dragonflies.

The Entomological Society had become embroiled in the day to day environmental impacts that affected butterfly or moth localities in such places as the Breckland, the New Forest, Joyden's Wood, Vert Wood and Dungeness. Many of these rich entomological sites listed by Charles Rothschild in his definitive report of the key sites in Britain (Rothschild 1915b) then became developed, for instance for forestry (Breckland) or for a power station (Dungeness). The Committee, and later the JCCBI sat through discussions over threats to the British countryside, which clearly had potential impact on butterflies. The aspirations of both Lord and Charles Rothschild for whole habitat conservation must have been shattered. It is interesting to read of the especially softened protests (from those on the Committee who were often either Captains, Majors, Lt-Cols or Colonels) by the Entomological Society to the War Office over potential damage to key sites. It was time for the Entomological Society to distance itself from the Committee (it already had kept at arm's length) and leave protection of the British countryside to the voluntary bodies or non-governmental organizations.

Historical background

REQUISITIONING

During the Second World War (1939-45) it was customary for the military to requisition land to do with what they wished (build installations, create training grounds, ranges, etc.). Many parts of the countryside were clearly of entomological interest especially heathlands, but there was very little that anyone could do about it if the MOD decided to pick a certain part of the countryside and carry out exercises. There was also the Dig for Victory programme which encouraged villagers and farmers alike to plough up all available land for the growing of crops. This was taken very seriously and millions of acres were put to the plough, including many ancient meadows. Thousands of butterfly sites disappeared, including, for instance, Adventurer's Fen - a noted part of Wicken Fen - which was lost forever. There was a War Agricultural Executive Committee appointed in each county to oversee the ploughing of all available land. After the war, some of this requisitioned land was returned to the owners, but much was not.

A spin-off of the SCPBI was the Nature Reserves Investigation Committee (NRIC) whose brief was to identify local nature reserves and designate them, all in the cause of nature preservation. It was set up by the Society for the Promotion of Nature Reserves (now The Wildlife Trusts) (Fitter, 1994). In 1943 a Conference on Nature Preservation in Post-War Reconstruction, chaired by the Rt. Hon. Lord Macmillan was written-up by the NRIC, and the results used as a basis for the site selection and acquisition of nature reserves nationwide. It was a comprehensive memorandum which outlined the sorts of reserves to be created, such as ones for individual species, and for different habitats. It remains a forthright policy document with fundamental principles (SPNR, 1943). The only lepidopterists on the panel of ten committee members were Captain C. Diver and Mr. J. C. F. Fryer. There is much in the report about the general public enjoying the countryside - reminiscent of the aims of The National Trust. It was amazing that the committee were concerned about the rape of the countryside as early as 1943. The worst in habitat loss, however had yet to come: 'The reservation (i.e. the concept of), before all suitable areas have been destroyed, of forests, heaths, downs, and marshlands, as well as of lakes, rivers, and the seashore, for the enjoyment of those who are compelled by circumstances to spend most of their lives in the man-made "deserts" resulting from urbanisation and industrial development, is a matter for urgent consideration.'

Some butterfly sites were still under apparent threat from the military in the years immediately after the war. One such site was the Isle of Purbeck in Dorset, of which a larger part was likely to be requisitioned. This upset a lot of people including Miriam Rothschild and E. B. Ford who wished that the Entomological Society would protest, write to the Prime Minister and encourage other societies

Historical background

to protest. 'Will you put it to the Council of the Ent. Soc. that we write a formal letter to the Prime Minister expressing dismay and astonishment at the requisitioning of Kingley Vale, Braunton Burrows, I. of Purbeck, Ashdown Forest etc. ...' (Rothschild 1946).

A Local Inquiry about the Isle of Purbeck was planned for 16 March 1948, and in April of that year Ford wrote to Riley; 'There are few ways in which the Society can contribute more to the cause of entomology, and towards saving the countryside, than by acting in this matter and supporting other Societies..... 'I should be sorry if the R. Ent. Soc. were to take the line now which it did when approached when the Report on the Conservation of Nature in England and Wales was being drafted....It then wished neither to submit evidence nor bring matters to the attention of the Committee. I do hope it will act now'. (Ford, 1948)

Ford was not to know then that land in the hands of the military was one of the safest protection policies for butterflies, as is now very clear 40 years on. Today the Isle of Purbeck, about 22,000 ha in size, is one of the major mosaics of habitat types in Dorset with heathland, chalk and grassland hills and many coastal sites; it also includes such important butterfly habitats as Lulworth Cove, Hartland Moor and Arne.

Today the RESL is represented by the JCCBI, and through its liaison with Wildlife Link which is a co-ordinating organization for all the many conservation organizations. The RESL also relies on advice from the Institute of Terrestrial Ecology (ITE) at Furzebrook, and from IUCN, both of whom took up responses to the British Government's commitment to the 1993 Biodiversity initiative. May (1993) drew attention to the relative interest in invertebrates including insects in his short but pithy review of British biodiversity,

In the last two decades there has been a phenomenal rise in enthusiasm for butterflies. This enthusiasm completely outstripped the initial interest in butterflies that the Royal Entomological Society had from its involvement with the Committee for the Protection of British Lepidoptera. There was a massive swing from a small group of professional entomologists to a mass growth area from the ranks of ordinary individuals who just liked butterflies. The British Butterfly Conservation Society (BBCS) was established on the 5 April 1968, by Thomas Frankland and Julian Gibbs, both keen amateur entomologists 'who felt that little has been done to co-ordinate or stimulate work for conservation of British Butterflies so far,' (BBCS, News Sheet, No.1. October 1968). Time was now ripe for a concerted effort at the conservation of Britain's butterflies.

MODERN CONSERVATION POLICIES

English Nature (EN), the current government body responsible for conservation, do not have a statutory obligation to conserve many British butterflies, but they are duty bound to look after the interests of the Large Blue and the Large Copper as they are protected by law.

There are however some conservators who cannot see why so much attention should have been focused on the Large Blue as a species that is at the edge of its range in England, when it still exists on the Continent. Britain and its wildlife are so much part of Europe that the presence or absence of species in Britain is only academic, not necessarily a point of complete unmitigated commitment. Yet, many resources have been poured into a few favoured species, arguably at the expense of scores of others.

Britain, as a series of islands on the west coast of Europe, must be considered in a European framework. British butterflies have populations which have adaptations suitable for the range of habitats influenced keenly by the typical Atlantic climate acting on an island. This shapes the uniqueness of the British butterflies and makes some English species slightly different, via their subspecies, from Continental examples of the same species. To the French, Britain is like an outdoor laboratory where extremes from the Continental norm can be seen.

It is from this European perspective that butterfly conservation policy should be shaped. The trouble is, this wider view has not been at the forefront of butterfly conservation in the past. There were those in office during the years of English Nature's predecessors, the Nature Conservancy Council (NCC) who pushed for a wider-based conservation policy for protecting habitats, and indeed the earliest responsible body, the Nature Conservancy (NC) was very much oriented towards the conservation of whole habitats. Before 1973 there was a lot of discussion on whole-habitat conservation. Indeed, the Large Blue emphasised the need to have detailed knowledge about the complex inter-relationships that exist in nature, so that the effects of changing methods of habitat management or the effects of pollutants on the environment could be monitored. This kind of philosophy has not changed. The Large Blue research is still lauded as a cure-all for general butterfly conservation.

The philosophy of the old Nature Conservancy (and the reason behind the acquisition of the National Nature Reserves (NNRs)) was to conserve prime habitats in an integrated way. Habitat Teams were set up in the late 1960s with whole-habitat conservation as a central aim, but there were research scientists in the NC at the time who were more interested in autecological studies (i.e. studies of individual species in relation to their environment).

All this came abruptly to an end with the Victor Rothschild 'customer-contractor' principle which started in 1973 and restricted the range of research

Historical background

allowed in science. The 1970s was a period of economic restraint brought about by an enormous leap in the price of fuel, which indirectly caused cash-strapped science politicians and advisers such as Rothschild to state that all future research should be at least of some economic value.

It was in 1973 that the Nature Conservancy was split into the Nature Conservancy Council and the Institute of Terrestrial Ecology (ITE). The reason why research today is centred on particular species, and not on the general health of organisms, or habitat research, is because of the way research now has to operate. Research contracts for the NCC (now English Nature) are increasingly specific and short-term in order to satisfy the administrative constraints of contract research. This is not always a good thing for habitats or butterflies, but it is the only way that this kind of research can operate. There were people within the NCC who strove to keep a modicum of whole-habitat research going but their views did not prevail. The result was increased contracts for studying individual species, including the Large Blue.

There was, and still is a need for a broader research programme which uses the Lepidoptera as a group of organisms to measure the changes taking place in British habitats and organisms over relatively long time-scales, though some would disagree. A further understanding of butterflies within habitats needs to be instigated to give a better perception of habitat conservation. The threat of global warming has at last brought scientists, mostly botanists, out into the fields to assess what is going on. Few lepidopterists have ventured along this productive path to find out how and if butterflies might profit by higher temperatures to give us a comparative handle on habitat change.

There was much debate at the time of the demise of the Large Blue that the central policy of the NCC was wrong. There was a call not to put all Britain's pecuniary resources into one species, since that might not bring results. After all there were 22,500 insect species in Britain, many of them rarer than the Large Blue. This was a powerful argument, often debated. A revolution was meant to be happening within conservation bodies. Whole habitats were to be conserved, complete with their own complement of species many of which were common, for it is the common species of today which may become the rarities of tomorrow. How things have changed. The new Species Recovery Programme (SRP) has returned to species protection. Twenty-two organisms (plants and animals) were targeted for 1992, only a few of them butterflies. Whole habitat conservation will have to be found within the framework of single species conservation, following the tenet that getting the habitat right for one species will inadvertently provide good conditions for thousands of others.

Historical background

FINANCIAL RESTRAINT AND A SINGLE SPECIES POLICY

Conserving butterflies is the least of any government's considerations. Butterfly conservation in Britain in its early stages was sponsored mostly by Lord Rothschild and the CPBL. From the 1960s onwards butterfly conservation has been financed by government money channelled first of all through the Nature Conservancy then the Nature Conservancy Council and more recently through English Nature, though it has been incredibly modest. It is true to say that the progress of butterfly conservation in Britain has been entirely determined by what monies have been available for projects. Despite this government sponsorship of butterflies conservation, supplemented by monies from the private sector, less than a third of British butterflies have enjoyed serious conservation treatment. Many of them have just had in-depth autecological studies carried out on them, which not always consider conservation.

Butterfly conservation carried out on behalf of Statutory authorities (such as the EN) costs money; but most of the conservation of butterflies in Britain, regarding everyday common species, is done through the voluntary sector, especially through the local wildlife trusts and Butterfly Conservation local groups, which involves a relatively small imput of money. There has been a history of targeting rare species, often ones which become extinct, with money from the state (cf. Stubbs, 1994).

The purpose of this final section is to assess the amount of monies provided for butterfly conservation so that one can see how and where the taxpayer's money has been spent so that a measure of efficiency can be assessed. It also provides an insight into what research is being carried out on which species of butterfly and how conservation research is progressing.

There has been a history of injection of funds into research just a few target butterflies in Britain, starting with the Large Blue butterfly and continuing with the Heath Fritillary, the Chequered Skipper and finally the larger fritillaries (Table 2.1). It was only in the 1990s that wider conservation considerations were funded such as formulating conservation strategies as carried out by Martin Warren.

The money expended on the Large Blue has been colossal. No other insect has had so much money attributed for it, even when extinct. If we accept that it is good and proper to reinstate the Large Blue from continental stock, then it still begs the question as to whether the money has been well spent. There can be only a handful of people who can see this butterfly each year in Britain, 20 years after re-establishment plans were put in operation.

The precise amount of money provided on the Large Blue project is almost impossible to calculate, since in the early years of the NCC, when indeed they actually did publish details of projects (which they did not for 1974-76 or for

Historical background

1977-88 which are crucial years in the extinction of the Large Blue) grants were given for all-embracing subjects such as 'recording data on individual species'. Post-extinction there have been at least £30,000 worth of grants apportioned to the Large Blue project. And up to 1993 over £70,000 have been given to unspecified 'conservation priorities'. It is heartening that in 1994 much larger grants have been made available, for instance with a substantial grant for an understanding of the metapopulations (i.e. inter-linked populations) of rare butterflies.

There are other butterfly projects of a conservation nature that have benefited from government's monies for butterfly conservation, notably the Butterfly Monitoring Scheme (BMS). Ernest Pollard pioneered the BMS which feeds back information on the state of butterflies during the breeding season using a transect method of observation. Others have had special interests, not always funded so well, such as Eric Duffey on the Swallowtail and Large Copper.

Another obvious feature from Table 2.1 is that only rare or endangered species have been targets for research. That is unfortunately where most conservators aim their money. The potential futility of this is manifest. The Large Blue was targeted. It became extinct. The Heath Fritillary was targeted, but so far it has held its own. The three big fritillaries are now targeted, whilst their populations are declining radically and their fate hangs in the balance. It seems we have learnt the art of leaving conservation too late. Another obvious omission in Table 2.1 is that there are now whole habitat surveys, of suitable ecosystems which are big enough to harbour sufficient common species to survive against all the odds.

Butterfly research has not had a history of over-generous financial support from the statutory authorities. Researchers involved in bird research have regularly received much more money in the form of grants than those who have been involved in research into butterflies and moths. In some years researchers into birds have received more money in a single year than has ever been given for butterfly conservation research since the emergence of the Nature Conservancy. For instance for the financial year 1988-89 the British Trust for Ornithology received £265,703 for 'Services in Ornithology' (NCC, 1989). This is more funds in one year, than all government funds for butterfly research as listed in Table 2.1. The reasons for this disparity lie somewhere in the higher levels of EN where proportionality of funding for the different disciplines is presumably fought over, for the authorities at EN are not prepared to pass any judgement on the subject. The lack of funding for butterfly research is unfortunate because butterflies are important ecological indicators of healthy environments, with a potential role to play in studying global warming; this potential has yet to be explored fully.

Historical background

Table 2.1 Nature Conservancy Council's, English Nature's and National Environmental Research Council's contracts on butterfly conservation, 1974-1994

Year, project, recipient and funds made available

1974-1975 NCC, 1st Report
no details of financed projects published

1975-1976 NCC, 2nd Report
No details of financed projects published

1976-1977 NCC, 3rd Report
Recording data on individual species: NERC(ITE) : £65,688
Site management information system: NERC(ITE) : £5,460 (4 years to 31 March 1980)
Autecology of the Large Blue butterfly: NERC(ITE) : £14,187 with the following:
Methods for monitoring population changes: NERC(ITE) : £14,647 (5 years to 31 March 1981)

1977-1978 NCC, 4th Report
no details of financed projects published

1978-1979 NCC,5th Report
Survey of butterflies of Bernwood Forest: Peachey, C. : £1,388
Large Blue management research: NERC(ITE) : £5,468

1979-1980 NCC,6th Report
Research of data on individual species: NERC(ITE) : £39,650

1980-1981 NCC,7th Report
Recording of data on individual species: NERC(ITE) : £35,000 (6 years to 31 October 1980)
Methods of monitoring population changes: NERC(ITE) : £20,760
(8 years, 3 months 31 March 1983)

1981-1982 NCC, 8th Report
Production of leaflet on butterfly conservation: BBCS : £1,800
Recording of data on individual species: NERC(ITE) : £40,000 (7 years to 31 March 1983)
Collation and assessment of site data for invertebrates: Hadley *et al.* :£40,000
Butterfly monitoring: NERC(ITE) : £7,000 (5 years to 31 March 1986)

Table 2.1 cont..

Historical background

1982-1983 NCC, 9th Report
Collation and assessment of site data for invertebrates: Hadley *et al.* : £37,009
(4 yrs to 31 Oct 84)
Recording of data on individual species: NERC(ITE) : £43,000 (3 years to 31 March 1985)
Autecology of Heath Fritillary: Warren, M. : £7,503 (3 years to 31 March 1985)

1983-1984 NCC, 10th Report
Collation and asses/t of site data for invertebrates: Hadley & Steele :£27,200
Entomological surveys by specialists: Hadley & Steele :£5,080
Autecology of Heath Fritillary: Warren, M. :£8,200
NCC to give £4,000 p.a. to Large Blue Committee (Stubbs letter to Kennard, 2 October 1984)

1984-1985 NCC, 11th Report
Collation and assess. of site data for invertebrates for Invertebrate Site Register
Foster *et al* :£32,029 (for 8 years to 31 March 1988)
Entomological surveys by specialists :£2,978 (to 31 March 1985)
Butterfly monitoring: NERC(ITE) :£8,190 (5 years to 31 March 1986)
Recording of data on individual species by BRC: NERC(ITE) : £47,956 (4 yrs to 31 March 1986)
Autecology of heath fritillary: Warren, M. : £8,810 (3 years to 31 March 1985)

1985-1986 NCC, 12th Report
Recording of data on individual species by BRC: NERC(ITE) : £62,380 (no time scales given)
Butterfly monitoring: NERC(ITE) :£12,000
Invertebrate Site Survey: Ball et al. :£53,930
Review of butterfly conservation priorities in S. England: Warren, M. :£12,560
Management of woodland for butterflies and other insects: NERC(ITE) :£30,500
Ecology of woodland fritillary butterflies: NERC(ITE) :£16,000

1986-1987 NCC, 13th Report
Butterfly monitoring: NERC(ITE) : £9,000
Invertebrate Site Register: Ball et al. :£65,950
Entomological surveys by specialists: various :£4,340
Review of butterfly conservation priorities in S. England: Warren, M. :£14,850
Management of woodland for butterflies: NERC(ITE) :£31,450
Ecology of woodland fritillaries: NERC(ITE) :£12,600
Effects of butterfly populations of pesticide-free headlands: Game Conservancy :£9,500

1987-1988 NCC, 14th Report
Butterfly monitoring: NERC(ITE) :£16,000
Review of butterfly conservation priorities: Warren, M. :£15,750

Table 2.1 cont..

Historical background

Ecology of woodland fritillary butterflies: NERC(ITE) : £16,800

1988-1989 NCC,15th Report
Butterfly monitoring: NERC(ITE) & EN : £16,000
Invertebrate Site Register: Falk *et al.* : £36,410
Entomological surveys by specialists: various : £2,137
Review of butterfly conservation priorities: Warren, M. : £19,622
Ecology of woodland fritillary butterflies: NERC(ITE) : £12,600
Management guidelines for butterflies... in plantations..: NERC(ITE) : £17,163
Ecology of the Chequered Skipper butterfly: Aberdeen University: £19,700
Effects of butterfly populations of pesticide-free headlands: Game Conservancy : £3,500

1989-1990 NCC,16th Report (from 1 April 1989 to 31 March 1990)
Land purchase at Park Corner Heath, E. Sussex: BBCS :£3,333
Butterfly monitoring: NERC(ITE) & Pollard, E. : £16,000
Ecology of the Chequered Skipper: Aberdeen University : £19,449
Developing butterfly conservation in Britain: Warren, M. : £4,747
Effects of butterfly populations of pesticide-free headlands: Game Conservancy : £2,625

1990-1991 NCC,17th Report (from 1 April 1990 to 31 March 1991)
Butterfly monitoring: NERC(ITE) and E.Pollard: £16,000
Ecology of the Chequered Skipper: Aberdeen University : £17,778
Developing butterfly conservation in Britain: Warren, M. : £11,001

1990-1991 English Nature, 1st Report
Autecology of the Chequered Skipper: Lean, I (Aberdeen) : £ 4,000
Butterfly conservation in Britain: Warren, M. : £12,800
Butterfly survey -Walton Crag: Skelcher,G.: £500
High Brown Fritillary, Dartmoor National Park: Oates, M. : £500
Large Blue butterfly: NERC(ITE) : £11,767

1992-1993 English Nature, 2nd Report
Conservation management of cereal field boundaries: Oxford University : £15,651.54
Monitoring of grassland transplant: Chris Blandford Associates: £2,017.81
Genetics of Endangered insect populations: Pullin, A. : £1,322
Autecology and conservation of the higher brown fritillaries: Warren, M. : £14,982
Large Blue butterfly, Species Recovery Programme: NERC (ITE) :£15,900
Large Copper butterfly, Species Recovery Programme: Keele University :£595

Table 2.1 cont..

Historical background

<u>1993-1994 English Nature, 3rd Report</u>
Brown Argus survey in Warwickshire & Worcestershire: BBCS : £750
Autecology & conservation of the High Brown Fritillary: Warren, M. : £16,000
Large Blue butterfly, Species Recovery Programme: Thomas, J. : £15,400

<u>1994</u>
Effects of population fragmentation on evolution of metapopulations of rare butterflies.
        Thomas, C.D. Birmingham University : £167,429 (NERC News, April, 1994)

Table 2.1 Notes: The ITE won contracts for 'recording data on individual species' and it is not at all clear how much of this was spent on collecting data on butterflies. In many years when research on Large Blues was progressing there is no specific mention of monies being available for Large Blue research, and one must presume that this came out of their catch-all contract work.

Sources: Mostly from annual reports of the NCC, EN and NERC.

Historical background

Notes

1. Lord Rothschild is Lionel Walter Rothschild (1868-1937); Nathaniel Charles Rothschild (1877-1923) is usually referred to as Charles Rothschild (father of Miriam Rothschild) (Rothschild, 1983).

2. The SPNR became the RSNC which became the RSNC, The Wildlife Trusts Partnership, which in 1994 became the Wildlife Trusts. This currently includes 47 Wildlife Trusts, 47 Urban Wildlife Groups and WATCH, its junior wing.

3. The CCCPN (which quickly became the BCCPN) set up the CPBL with the help of the RES in 1925, which then had a wider remit for the British insects and was called the SCPBI. This then became the JCCBI. The 'Insects' of the JCCBI became the 'Invertebrates' at the 55th Meeting on 30th October, 1992. The JCCBI has had a history of widening its own awareness, from Lepidoptera, through Insects to Invertebrates.   In 1942 the Conference on Nature Preservation in Post-war Reconstruction, established a committee to draw up proposals on the establishment of nature reserves. Typical butterflies to have taken much of the time of these committee meetings include the Large Blue and Large Copper. The well-being and status of these, particularly the Large Blue, have been the subject of discussion of at least 12 committees.

4. Mr. Newman was Leonard Woods Newman whose son was L. Hugh Newman (Gardiner, 1993).

5. Miriam Rothschild states that when her father died, Professor Tansley said there was noone else to continue the enthusiasm (Miriam Rothschild, pers. comm. 1994).

6. H. M. Edelsten (1877-1959) was an important lepidopterist employed at the NHM where he arranged the collections (and is noted for his discovery of the life cycle of the Marsh moth, Athetis pallustris); N. D. Riley was Keeper of Entomology and was initially involved with the SPNR, later to become well known through his 'Higgins & Riley' book; J. C. F. Fryer (1886-1948) was later to become Sir John Fryer, noted expert on papilios of Sri Lanka, and W. G. Sheldon (1859-1943) was a noted microlepidopterist.

CHAPTER THREE
Why conserve butterflies?

...'conservation in the modern sense of the word ceased to exist in the United Kingdom.' .. after the death of the Founder of Nature Conservation, Charles Rothschild   Miriam Rothschild, 1979

The answer to the question of why conserve butterflies is very simple. We have only 59 breeding species, and all are key species. In addition we have migrants which can come and go as they please, their arrival in Britain often relys on successful propagation in other countries; we can only give them sanctuary, or a better sanctuary once they are here.

Butterflies are part of Britain's diverse faunal list, and a colourful one. They might be small in number of species, but their allure and their signals to the general public as examples of the living world quite outstrip their puny numbers. Small Tortoiseshells, Peacocks and Red Admirals do more for promoting butterflies in the urban venue, and are great ambassadors for butterflies in

Why conserve butterflies?

general. Think of just how many commercial enterprises use the butterfly logo for promoting their services.

To many people (not butterfly enthusiasts) butterflies are just butterflies, with no particularly attractive features, which are simply lumped together as 'bugs'. To them butterflies are rarely seen, only perhaps as a pest in the cabbage patch, and are not regarded as worth conserving. Nature programmes on television however have helped to change the appreciation that butterflies have in the eyes of the public, although the loss of the Chequered Skipper from England passed without the general public getting upset about its demise. In contrast the Large Blue was the subject of short-lived national acclaim when it became extinct in 1979.

One of the major considerations in answer to the question why conserve butterflies is that they are disappearing fast. Their beauty and their role at a local level of indicating the quality of a habitat combine with our knowledge of their overall demise to create an urgent need for their conservation. The facts are clear, spelt out by The Nature Conservancy Council as long ago as 1977, when they said that 95% of butterflies would be lost from the agricultural landscape if the pressures that farmers put on the land continue (NCC, 1977). These facts were expanded upon by Marion Shoard in her candid books on the state of the countryside in 1980 and 1987.

SPECIES RICHNESS AND DIVERSITY

The terms 'species richness' and 'diversity' need some explanation since they are frequently used in talking about popular butterfly localities. In a sense they are synonymous terms since one dictionary definition of diversity is ...'the species richness of a community', but academics have different ways of separating them. Measures of diversity often take into account not only the number of species but also the number of individuals in each of the species in an area. Lay people like to talk about the number of butterflies found in a garden, a nature reserve, or on the county records, and all this is a measure of the richness of that habitat. What is most important is the type of habitat and the diversity of the habitat types in a given area.

Butterfly conservationists like to describe a certain locality or reserve by the number of butterfly species it contains, and a representative sample of such sites are listed in Table 3.1. There is a certain amount of satisfaction about having a high number of butterfly species on a site, although the implication that one reserve is better than another is not valid since sites are not the same. Some habitats are simply better than others for butterflies, for example woodland glades, chalk grassland, or urban habitats are better than beechwoods or bogs.

Why conserve butterflies?

Bernwood Forest in Oxfordshire has always been a classic place for butterflies and other insects, and it boasts more species than most localities in Britain (Table 3.1). Caroline Peachey, who did a lot of observational research on the butterflies of Bernwood, recorded 41 species between 1975-1979; 45 species have been recorded over the last century (Peachey, 1980). To this figure could be added another, that of a possible sighting of a Large Tortoiseshell. Bernwood Forest is made up of many parts, and in the late 1970s there were 40 species recorded from the Shabbington Complex and 30 from Bernwood Forest. Peachey accounted the particular richness of Bernwood Forest as being due to a combination of factors including it geographical location, size and past and present management; it being a remnant of a much larger forest.

Bernwood Forest vies for prime position with Ashton Wold, a Site of Special Scientific Interest (SSSI) situated 2 miles (3.2 km) from Oundle, which has had 46 species of butterflies recorded from the area (Rothschild, 1975, 1989). There are probably two other woodlands in Britain with an equal number (Miriam Rothschild, pers. comm. 1994). From the publications of Bree (1852) and Morris (1895) and the work of N. C. Rothschild it is possible to gain some idea of this fluctuation of numbers at Ashton Wold over 140 years. For instance the Black-veined White is now extinct. In 1851 the Comma was abundant at Ashton but totally absent between 1900-1921. The first specimen seen or captured (after a lapse of 21 years) was in 1922 and during the next year or two this species became exceedingly numerous and is still well represented. Both Bernwood and Ashton Wold now share a serious decline in the different species present and the number of specimens on the wing. One is fearful that present day conditions will no longer permit a long-term recovery. Nevertheless after an absence of 30 years a pair of Silver-washed Fritillaries re-appeared in Ashton Wold in 1992 (Rothschild, 1992). Moreover the introduction of a mixture of wild flowers to the 'set-aside' grass fields alongside the woodland has resulted in a dramatic increase in the number of browns and the Common Blue, which had been dwindling steadily in numbers during the last 20 years, and the Grizzled and Dingy Skippers and Green Hairstreak are also back.

The loss of certain species at Ashton Wold such as the Marbled White, Grizzled and Dingy Skippers was undoubtedly caused by the destruction of their natural habitat by order of the Ministry of Agriculture (during the Second World War) which required cereal crops to replace the rough grassland and short turf. Furthermore after the death of Charles Rothschild in 1923 silviculture was abandoned and lack of labour during and after the Second World War resulted in a marked increase in shade and the loss of open ridings in the wood. The sudden disappearance of the Chequered Skipper - abundant in several areas in 1952 - remains a mystery since its habitat was not interfered with. It is also strange that the Duke of Burgundy Fritillary vanished. On the other hand Miriam Rothschild

Why conserve butterflies?

has recorded an increase in numbers of four species at Ashton Wold, the Holly Blue, the Speckled Wood, the Gatekeeper and the Small Skipper (Rothschild, pers. comm. 1994). But she feels it is extremely unlikely that modern agricultural methods, the use of toxic chemicals, the reduced woodlands and the overgrazing of the plants beneath the trees by 'wild' deer will ever permit a return of the 'good old days' when Bree (1852) watched Marbled Whites and Meadow Browns chasing Purple Emperors along the ridings in Ashton Wold. The entomological trophy of this district was, in days gone by, the Large Blue which flourished at Barnwell Wold and one specimen was recorded from the outskirts of Ashton Wold. Lord Hillingdon was the last entomologist to see a live specimen of this butterfly in Northamptonshire. In 1940 he told Miriam Rothschild he had a good view of one on rough ground near the Earl of Spencer's property. It would be of great interest to look at his large collection but since his death its whereabouts has remained a mystery.

The diversity of butterflies (or of any other group of plants or animals) can be measured quantitatively in any habitat, using a variety of different, and variously complicated measures. Those who wish to pursue these mathematical formulae and models are directed to Southwood's (1975) book. Suffice it to say that as any open habitat matures the vegetation changes to a relatively stable climax community, such as mature woodland, the fewer butterflies are likely to be found there, since butterflies generally are expert exploiters of the early stages of plant succession. Any habitat, for example chalk grassland, would develop into woodland if it were not grazed. But perhaps the ordinary level to which everyday lepidopterists wish to pursue diversity studies is through the work of the Butterfly Monitoring Scheme run by Ernest Pollard which has enormous popular appeal, and which produces useful quantitative data.

BIODIVERSITY

The biodiversity debate puts diversity into the international frame. It is not enough to think as an anglophile or a europhile, the international perspective must be embraced to get a better picture, an overview of British butterfly conservation on the international stage. Such were the inspirations, aspirations and much political hype emanating from Latin America with the Rio Conference in June 1992.

Governments around the world, including the British government, signed the Rio convention, committing themselves to maintaining biodiversity. Britain laid the foundation for conserving UK biodiversity by establishing the Darwin Initiative. Comments and suggestions were sought during the summer 1993 from

Why conserve butterflies?

Table 3.1 Species richness

Localities	Number of species	Reference
Ashton Wold Nature Reserve SSSI, Northants.	46 (1920-1974)	Rothschild, 1975,1987
Bernwood Forest SSSI, Oxfordshire	45 (1880-1980)	Peachey, 1980
Bernwood Forest SSSI, Oxfordshire	39	NCC, 1983, page 80
Ministry of Defence, NW Hampshire	44	Yorke, 1987 News 39:31
Wood Walton Fen NNR	38	Morris, 1967a
Park Corner, East Sussex	36	Tatham, 1990 News 44:9
Park Corner, East Sussex	37	Hicks, 1990 News 45:28-29
Bill Smyllie Reserve, Gloucestershire	35 (locally)	Roper, 1993 News 53:32-33
Somerford Common, Wiltshire	35	Wiltshire Trust leaflet
Monks Wood NNR, Huntingdonshire	33 (all time = 41)	Morris, 1967a
Elliot/Swift Hill SSSI, Gloucestershire	29	Gloucestershire Wildlife Trust
Castor Hanglands NNR	32	Morris, 1967a
Llanymynech Rocks SSSI	30	M'gomeryshire & Shropshire Trusts
Monkwood, South, Worcestershire	31 over 8 years	Badham & Williams, 1986 News 37:27
Monkwood, South, Worcestershire	30 since 1978	Tatham, 1986 News 36 page 1
Finemere Wood, North Buckinghamshire	30 (about)	Glover, 1990 News 45:23-4
Pembrey Country Park, & Pembrey FNR	30	Carmarthenshire Trust
Loggerheads Country Park, Clwyd	29	Clwyd Trust (pers. comm.)
Cefn Yr Ogof, Clwyd	29+	Clwyd Trust information
Coedgelli-deg, South Wales	28	Anon., 1988 News 40:11
Magdalen Hill Down, Hampshire	27	Yorke, 1990 News 45:26-27
Farthinhoe LNR, Northamptonshire	26	Northamptonshire Trust
Marfold Quarry, Clwyd	26	Clwyd Trust (pers. comm.)
Y Graig, Clwyd	25	Clwyd Trust (pers. comm.)
Aberduna, Clwyd	25	Clwyd Trust (pers. comm.)
Prestwood Picnic Site LNR, Buckinghamshire	25	Roberts & Beaven, News 55:19
New Forest, Hampshire	25 (one day)	Yorke, 1990 News 45: 38
Twywell Quarry, Northamptonshire	24	French, 1989 News 43: 20-21
Collyweston Quarries, Northamptonshire	24	Northamptonshire Trust
Craig Fawr, Clwyd	24	Clwyd Trust (pers. comm.)
Fenns Moss NNR	24+	Clwyd Trust (pers. comm.)

'News' refers to Butterfly Conservation's News. See Acronyms for SSSI, FNRs, NNRs etc. The data presented here are not directly comparable, since some figures are 'all-time' records, 'seasonal' records, or found 'locally'; where this is know the information has been given. The number of species recorded from a site must be considered on its own merit. For instance a seemingly low number of species for the Twywell Quarry SSSI in Northamptonshire is relatively high for that county since only 32 species of butterfly have been recorded in the county in recent times. This is in contrast to the all-time record of Rothschild (1987) for Ashton.

Why conserve butterflies?

the learned bodies on how to progress with this positive intention. Groups such as the Linnean Society responded by soliciting comments from their Fellows; the Royal Entomological Society left the JCCBI and IUCN to reply on their behalf. In December 1993 the Government published its White Paper on Science and Technology (HMSO,1994) and the NGOs responded to this independently.

Of the ten proposals submitted by the Linnean Society to the Government, two of them would specifically favour butterflies: 5. More support for systematic biology research and databases and for basic taxonomy and curation and 6. Special funding for studies of complete endangered ecosystems leading to the formulation of conservation strategies for them (Linnean Society, 1993). It is good to see whole ecosystems being proposed for further research, rather than individual species. If monies were made available for those species that are scarce in Britain then butterflies would score very well. They deserve a bigger share of the conservation cake.

The British Government sought to reassure everyone that existing legislation, such as the Wildlife and Countryside Act 1981, was sufficient to rely upon to conserve the flora, fauna and habitats of the UK. The problem with existing legislation is that it does not do the job of protecting or conserving what it is meant to. The track record for legislative wildlife conservation indicates that it cannot be relied upon; for instance SSSIs are disappearing at about 5% per year along with their mosaic of inter-related flora and fauna including butterflies. Habitat loss is continuing unabated with few fine localities left for butterflies. Existing legislation just will not do.

Some of the leading NGOs seized the opportunity to put forward their own vision for the Darwin Initiative, which was published as the *Biodiversity Challenge, an Agenda for Conservation in the UK* (RSPB, 1994). The primary authors of this report were the Royal Society for the Protection of Birds (RSPB), the Royal Society for Nature Conservation (RSNC), the World Wide Fund for Nature (WWF), Friends of the Earth (FOE) and Plantlife, but there were other Non-Governmental Organisations (NGOs) involved, bringing the collective number of people represented by these conservation groups who were concerned about the welfare of various parts of the countryside to about two million people. The major concern to see implemented through the Darwin Initiative was 'the prevention of extinction and maintenance of range' of the species conserved (Stubbs, 1994) and fourteen butterfly species were thus targeted for potential conservation (Table 3.2).

Why conserve butterflies?

Table 3.2   Butterflies targeted for conservation.

1. High Brown Fritillary (*Argynnis adippe*). Maintain populations at all occupied sites.
2. Pearl-bordered Fritillary (*Boloria euphrosyne*). Halt rapid decline - 63% extinction rate per decade in southern Britain. Restore to former range with at least three sites per occupied county.
3. Chequered Skipper (*Carterocephalus palaemon*). Extinct in England. Further clarify range in Scotland, currently 24km squares known, and then at least maintain range extent by ensuring correct management of 20 core metapopulations. Re-establish in six English sites.
4. Large Heath (*Coenonympha pamphila*). Maintain all lowland raised bog populations. This species is known to be genetically very variable with clines and three sub-species in the UK.
5. Marsh Fritillary (*Eurodryas aurinia*). Maintain range and safeguard at least 10 clusters of sites with bio-geographic representation.
6. Silver-spotted Skipper (*Hesperia comma*). Halt decline and restore range.
7. Large Copper (*Lycaena dispar*). Reintroduced to Britain. Re-establish group of interconnected populations in eastern England.
8. Large Blue (*Maculinea arion*). Re-established in Britain, extend to eight former sites in England.
9. Glanville Fritillary (*Melitaea cinxia*). Maintain present range on the Isle of Wight, recognising the linkage necessary to maintain metapopulations.
10. Heath Fritillary (*Mellicta athalia*). Halt decline of populations in Kent and restore its 1980 status, 25 colonies within a metapopulation. Maintain range in Cornwall, Devon and Somerset. Maintain re-established population in Essex.
11. Large Tortoiseshell (*Nymphalis polychloros*). Possibly extinct for about 15 years. Maintain any existing populations if discovered or re-establish if feasible.
12. Swallowtail (*Papilio machaon britannicus*). Restore range and re-establish at Wicken fen. On the Norfolk Broads increase area of occupied breeding habitat by 50%.
13. Black Hairstreak (Str*ymondia pruni*). Rapid decline halted by conservation programme. Maintain required management on all 30-35 sites, 12 of which are reserves. Increase habitat and population by 50% on at least six of the less secure sites.
14. Lulworth Skipper *(Thymelicus acteon)*. Maintain range.

Why conserve butterflies?

Targets were set by this group of NGOs for helping to sort out which of the many deserving species would be able to be conserved in the long term. Maintaining diversity of species and habitats is clearly their aim. One of their criteria is that endemic species would qualify for 'directly limited conservation resources'. This rules out all British butterfly species, but some subspecies of British butterflies might possibly be considered Any species which has in Britain a population that is internationally important in the European framework would qualify. Thus, things might get so bad for the Marsh Fritillary on the Continent that the British populations take on an increasingly important role. Marsh Fritillary populations might even disappear from much of the Continent, leaving Britain's surviving colonies in a better international context. Some butterfly species, however, have been targeted because their populations have declined to specific levels, for instance if 50% of their range has declined since 1960, or if they become restricted to fewer than 15 10-kilometre distribution squares (as indicated on the dot distribution maps). Another criterion under which a few species qualify is where they have become extinct since 1900! The NGOs also published a species action plan for the Large Blue (see Appendix 1).

With regard to the types of habitats that are deserving of conservation, the NGOs group came up with the following, all of which are very useful for supporting butterflies as well as other species: lowland heath, machair, lowland bogs, lowland wet and dry grassland, fens, carrs, eutrophic waters, native pinewoods, salt marshes and wood pasture and parkland. Each of these habitats is designated as being of 'high conservation importance' (the highest level) for the insects which they harbour. In January 1994 the UK government published their own UK Action Plan on Biodiversity (HMSO, 1994). This is a multi-authored document which presents a masterly overview of the current status of the flora, fauna and habitats of the UK, but that is as far as it goes.

Because there are so few species it makes it doubly more important that butterflies are conserved, all 59 species. Each species from a Cabbage White to a Dark Green Fritillary would ideally carry equal weight, but not all are in real need of urgent conservation action so priorities have to be established.

CONSERVATION: THE LAST CHAPTER?

It is only in recent years that we have really woken up to conservation. The subject of 'conservation' hardly figures in butterfly books prior to the 1980s, and if mentioned was confined to an afterthought, a concession to the new way of thinking, often as a last chapter to a book, even in my own *The Natural History of Butterflies* (Feltwell, 1986) and in Paul Whalley's landmark book *Butterfly Watching* (Whalley, 1980).

Why conserve butterflies?

E.B. Ford's classic book *Butterflies* published in 1945 gives us nothing on the subject, merely a few comments on the natural vagaries of butterfly populations, and on collecting. Ford plays down any adverse effects of over-collecting although he acknowledges that sometimes collectors did exterminate species: 'it is my opinion that over-collecting, itself an obvious evil, has not often been responsible for the destruction of butterflies, and that.. the danger has been unduly stressed.' The conservation of the species revolved around the virtues of collecting then.

It was not my intention to include anything other than a few words on collecting in this book, but since it seems to have permeated every facet of the conservation of butterflies I found myself irresistibly drawn into discussing it (see Chapter 11). The term 'over-collecting' speaks for itself and needs no further comment other than there can hardly be any qualification or mitigating circumstances for it. If 'collecting' was the buzz-word of the moment earlier in the twentieth century, then 'conservation' is the buzz-word today. It is a pity that most of the early lepidopterists were not conservationists (there were exceptions) and that conservation was not the rage then as it is now.

In recognizing that some species were 'systematically persecuted', and that this was a 'deplorable' event, E. B. Ford believed that the 'greatest danger to British butterflies' was through indirect effects such as habitat loss - something that most would agree with today: 'No woodland species can survive the destruction of our forests or the awful change from deciduous trees to conifers, while the conversion of moors and wastes into third-rate agricultural land is not always a complete compensation for the loss of natural beauty and of wild-life which is sustained in the process.' (page 143).

Ford appears never used the word conservation. He busied himself to giving an accurate health check of the British butterflies. His book was, and still is, simply a good read for butterfly enthusiasts, to understand the range of the British butterflies and how they were faring. The 1940s were still a period of discovery, and the extinction list had not really sunk in. There was therefore no reason to be concerned; the clover fields, lucerne fields, cabbage fields and kitchen gardens were still bobbing with a myriad of butterflies (mostly white and yellow) and the lanes were full of a wide variety of browns and skippers, for this was the period before the damaging organic chemicals were let loose on the environment (they became increasingly available after the Second World War in the late 1940s). This was a period of butterfly discovery, and the Ford book set the scene for this particular enthusiasm.

The loss of favourite haunts of butterflies was not lost on Ford however. The Forestry Commission (FC) was established in 1917, some 25 years before the publication of Ford's book, in an attempt to make Britain self-sufficient in timber (it has not succeeded in this endeavour today), and some of those excruciatingly

Why conserve butterflies?

tempting woodland haunts of the Victorians were cut down and planted with conifers.

Of course the study of butterflies was one of original enquiry into how they worked, what their genetics were and a general appreciation of butterflies as living things. There was no apparent urge to integrate them in the environment - indeed, this relevant word was also left out. Lepidopterists were in full swing in their enthusiasm for collecting, and boasting about the best localities to find butterflies. Newman's book on the *Butterfly Farmer* strongly emphasizes the widespread collecting for sale that went on in Britain, especially the industry set up around collecting that his father set up in Bexley (south-east London) and one can see adverts of this reproduced in his book.

Conservation may well have been the 'last chapter' in some books since it made concessions to the demands of the new conservation lobby; however, conservation is not, prophetically, the 'last chapter' in the demise of Britain's butterflies. Let us hope the epitaph of butterflies is not presumed by this previous back-seat status. I trust that in this book, at least, butterfly conservation is the major subject from page 1 to the end - the first book ever to deal with British butterfly conservation throughout - and that it will stimulate others to investigate further.

ROLES OF BUTTERFLIES IN CONSERVATION

In conserving butterflies in the countryside it is necessary to know as much as you can about the role they play in the habitats in which you find them. This role in nature is called a *niche*. Every living thing has a place in nature and that role reflects the integration of that plant or animal in habitat.

The role of the butterfly in the environment is a difficult concept for planners to use when making legislative conservation measures, or policy for butterfly conservation. Those who conserve and legislate for the conservation of butterflies can have difficulty in categorizing butterflies to particular habitats. It is easy enough to draft legislation for a particular habitat, but not all butterflies are exclusive to those habitats There are though, butterflies that are severely limited to their habitats, and these are therefore easier to legislate for, but not necessarily easier to conserve.

Many species of butterfly, however, live in habitat mosaics, that is, they are various parts of different, and often adjacent habitats. Obvious examples are the larger woodland fritillaries which use a mosaic of the woodland itself, woodland margins and open scrubby areas. In this case the mosaic of different habitats is a dynamic system which is always changing. Many habitats are dynamic in that the vegetation changes over time for example scrub will invade ungrazed grassland. Climax communities (e.g. mature oakwood, mature beechwood etc) in contrast, remain the same for long periods. Managing butterflies in these transitory and dynamic habitats, or in a mosaic of dynamic habitats, is a nightmare to any conservator of butterflies. The 'do nothing approach' so often practised, will not always work, and the temptation is to manage the habitat or mosaic in order the keep the habitat as it is, or as it was found for the particular butterfly which is being conserved there.

It might seem odd to some people but the Ministry of Defence (MOD) is one of the leaders in butterfly conservation today. They have done more, unwittingly, for nature conservation in the long term than other large conservators of the countryside. It is now a well established fact that certain tracts of land, especially those in Wiltshire and Dorset have more butterflies than other comparable places elsewhere. The MOD have come a long way since they were 'written-to' by members of the *Standing Committee for the Protection of British Insects (SCPBI)* for making use of various interesting localities for their manoeuvres, and a little later for requisitioning lands during and after the Second World War. If members of the SCPBI knew then that butterflies would eventually be superbly conserved by the MOD in the long term, perhaps they would not have protested in the first place.

By 1983 the MOD held about 242,820 ha in the UK, which made them the third largest landowner (Walker, 1983). The beauty of their holdings was that they contained a good cross section of suitable butterfly habitats such as chalk downland, heathland, woods, bogs, marshes and 341 km of coastline with cliffs and extensive dune systems. Under the leadership of Lt Col C. N. Clayden (Retd) there were 170 conservation groups in the MOD in 1983 involving 4000 people, many of whom were fascinated with butterflies. The MOD's little bulletin called *Sanctuary*, finally made a giant leap to a glossy A4 magazine in 1988 bringing colour images of butterflies onto the pages. The MOD are therefore a major force in the conservation of butterflies in Britain, and have gone to great lengths to foster enthusiasm for butterflies as symbols of conservation generally.

CHAPTER FOUR
Flagship species

'**Our country as we know it is vanishing before our eyes. We see its coasts and fields chosen as sites for houses and power stations. Its green hills sprout steel masts and pylons, its old oakwoods are clear felled.**'... Sir William Wilkinson, 1990, from a Nature Conservancy booklet of the 1950s.

CHARACTERISTICS OF 'FLAGSHIP' SPECIES

A flagship species is one that draws peoples' attention to the wider issues of the conservation of butterflies (or other group of organisms); they are also called keystone species. It has to be said that certain flagship species make good news stories, none so good as the Large Blue butterfly, especially when it became extinct in Britain.

The English Swallowtail is a fine example of a flagship species because it is big, easily identifiable and spectacular. All these characteristics are excellent for the

Flagship species

general public who are able to associate this butterfly with a particular region, the Broads, a major tourist and recreation area in Britain. The Swallowtail is certainly a flagship species used extensively by the tourist industry precisely because it pin-points the area they want to promote. Although the Swallowtail is singled out for its outstanding qualities, its flagship status does much to further interest in butterflies in general.

There are other flagship species too, such as the Large Copper, which has taxed the minds of conservationists since the 1920s to the present day, and the Black-veined White which became extinct in 1925 and which some lepidopterists hanker to re-introduce. The Chequered Skipper enjoyed some flagship status when it became extinct in England in 1975, but the Mazarine Blue which became extinct 1906-1920 has never really attracted a lot of attention, so extinction is no sure path to flagship status. Local extinctions hardly get any flagship status at all, only, if they are lucky a mention in the press, as did the chalk downland race, *cretaceus* of the Silver-studded Blue when it appeared to have become extinct by 1981 (NCC, 1981, cf. Birkett, 1995). This species has never hit the headlines. Nor has the apparent extinction of the Common Blue from the Shetlands come to the public attention (Arion, 1993). What does a butterfly have to do to be selected? They all need sponsors, and to catch the imagination of the general public.

One is bound to question whether the inadvertent 'sacrifice' of the Large Blue, which became a flagship species, was justified for the sake of the remainder. Has the cause of butterfly conservation been furthered by its demise? Yes, conservation has been furthered, but a little too late. If either the Heath Fritillary or the Chequered Skipper become extinct in Britain we will really have to re-examine fundamental conservation strategy.

It is certain that there are plenty of other insects that are unsuccessfully running the ecological gauntlet in the British countryside and whose demise goes unreported. There is an overall decline in the fortunes of many butterflies and it is unfortunate for the majority that only a few species hit the limelight.

Butterflies therefore carry the conservation flag for other species. One of the other roles of flagship species is that they draw attention to butterflies as indicators of the health of the countryside. They are 'conservation monitors'. A large number of different butterfly species in any type of habitat is a reasonable indicator that the flora is reasonably diverse too, with certain reservations, since butterflies are so dependent upon plants; in her studies in Bernwood Forest, Peachey found that butterflies used 51 species of plants just for sources of nectar sources (Peachey, 1980). It also indicates that the habitat has not been overly degraded or improved with agricultural chemicals. Here butterflies have been exceedingly good indicators of key habitats.

It seems that a certain degree of secrecy and intrigue has been regarded as a successful ingredient for a flagship species (cf. Pyle, 1981) something which is still perpetuated today, especially with the Large Blue. The re-establishment of an extinct butterfly makes a good news story, just so long as the locations are kept secret. Disclosing information on rare flagship species is clearly a very risky business, even in today's world of openness and freedoms. This covertness has at least a good pedigree since the Large Blue is still nurtured on a suite of eight sites some of which are known as 'X' and 'Y'; and there have also been exploits of a secret Mr. X.

Adams & Rose (1978) make the valid point that as a species becomes rarer, it is perceived as much more interesting. The Large Blue clearly became a much more interesting subject, and more valuable ecological indicator, just before it became extinct. Norman Moore (1987) put it cogently: 'The extinction of a handful of species such as the Large Blue Butterfly has done more to rouse public opinion than the production of all the statistics which demonstrate the massive declines of many surviving species, even though this is a far more important matter. Why are disasters so important a means of communication?'

He may well ask. Did we have to have the demise of the Large Blue to highlight the plight of the remaining butterflies? Conservationists need to be reminded not to be blinkered against the reality of the living world and all its integrated flora and fauna. The ecological world hardly revolves around single species and their conservation. We all know that, but still pursue single species conservation.

The latest focus of flagship attention has been the Marsh Fritillary of which Butterfly Conservation have made more people aware. The plight of the butterfly in Glamorgan, South Wales, where it is threatened with local extinction, is typical of many butterflies; this stimulated Butterfly Conservation to say that 'The loss of butterflies such as the Marsh Fritillary is a strong indication that all is not well with our countryside' (Butterfly Conservation Press Release, 30 September 1993).

More than 10 years earlier Jeremy Thomas was saying that 'over three-quarters of Britain's butterflies are experiencing some sort of a decline, that at least one quarter are in severe decline' (Thomas, 1981a). His prediction that there would be further national extinctions before the year 2000 is partially correct, and may well yet be true, when one considers the retreat westwards that some butterflies such as the larger fritillaries have experienced. As butterflies become rarer or locally extinct their flagship rating increases.

There have been various attempts over the years to whip up enthusiasm for butterflies, and one of the first was the Festival of Britain in 1951; this was a little clouded in controversy then as the Committee for the Protection of British Insects had to write to Mr. Newman asking him to withdraw or modify his adverts for fritillaries to be caught especially for the exhibition. How nice it would have been

Flagship species

if butterfly houses had been in existence then.

In more recent years there have been other butterfly promotion events such as Operation Butterfly, and National Nature Week, as well as the Brimstone Project set up by the Edinburgh Butterfly Farm to introduce larvae of the Brimstone into schools of the Lothian and Edinburgh region of Scotland. Some local entomologists became concerned about the release of Brimstones into an area where this did not occur naturally, and this was transmitted to the JCCBI for their comments, but none was given.

Perhaps the greatest awareness of the general public to the promotion of butterfly conservation was through 'Butterfly Year' in 1981 (see Thomas 1981a; *Daily Telegraph*, 16 December 1981). This focused attention on the plight of butterflies in Britain and was promoted by various scientific and conservation bodies. On 13 May 1981 four postage stamps were released by the Post Office, showing two rare species and two common ones, the Large Blue, Chequered Skipper, Small Tortoiseshell and Peacock. As the NCC stated in their 7th Report (NCC, 1981) Butterfly Year launched an appeal set up to raise funds to acquire reserves and support management projects. Butterfly Year was the first national campaign on insect conservation. In the same year Guernsey also issued a set of butterfly stamps, all of them showing common species: Common Blue, Red Admiral, Small Tortoiseshell and Wall.

Since Butterfly Year there has been Project Papillon in 1984. This was an endeavour to release large numbers of common butterflies onto Hampstead Heath during the summer of 1984. It was co-ordinated by David Lowe of Guernsey, a keen promoter of butterfly houses in Britain, and supported by the Guernsey Tourist Board. The nature conservation authorities did not take kindly to this release, for it is recorded that it had no conservation aim and that it was an entirely commercial venture (Minutes of the JCCBI, 6 December 1984). Paul Whalley was instructed to write to the Lieutenant Governor of Guernsey explaining the Committee's opposition to the scheme, but also saying that they would like to be consulted about any further contemplated releases.

Butterflies were featured as important subjects in Britain's gardens with the launch of the Flora for Fauna campaign on 19 October 1994. This was promoted by the Duchess of Hamilton to highlight the preferred plants for wildlife in Britain's gardens. Supported by many of the conservation bodies in Britain, it ran parallel with the European Conservation Year - ENCY '95, and, in one of its novel presentations, spread its message through the length of Britain on board a train. Flora for Fauna is interested in the welfare of many of the common butterflies of gardens, but there have been other British species which have captured the attention.

Flagship species

CHEQUERED SKIPPER

The Chequered Skipper became a flagship species in 1975 when it became extinct in England, catching butterfly conservators off their guard. The British population now resides in Scotland, holding its own there with attempts to re-establish it back into England, from Scottish stock. The last English Chequered Skipper was seen in Rutland in 1975, but it had also died out at one of its strongholds, Castor Hanglands NNR in 1971. This so alarmed the nature conservation authorities that a preliminary report, and a much belated one, on the status of the Chequered Skipper was rushed out by Lynne Farrell of the Nature Conservancy Council in September 1973. Farrell stressed throughout that the report is very much preliminary, and its assessed information on the species was gathered from all sources. There is no mention of methods of conservation (since none was used); the report tried however to get to the bottom of what had gone wrong.

The results were what one might have expected. It was easier to assess the situation in retrospect and to see how the intensification of agriculture contributed to the demise of the species. This was a familiar story, and the results were applicable to many other British species. The negative environmental impacts that Lynne Farrell collated for each of the former Chequered Skipper localities were as follows: clearing of habitat; engulfing of the habitat by surrounding land; fragmentation of the habitat; domination of the habitat by coarse grasses; autoscything to the detriment of larvae; burning of heath to the detriment of larvae and pupae; marginal land ploughed and the loss of 'fringe' habitats.

The trouble with any assessment like this is that it is subjective. It is qualitative, not quantitative, and critics would ask for the evidence. The truth is there is no empirical evidence. So one cannot lay the finger on any of the above environmental impacts and say they contributed to the demise of a certain species; although they were implicated is undoubted. It was only to reinforce the possible impact of agricultural intensification and widespread use of chemicals that T. R. E. Southwood mentioned this in the Royal Entomological Society of London's statement on Insect Conservation in 1965.

Much of the early research on the Chequered Skipper was done by Ray Collier from the early 1960s when it was abundant in places like Castor Hanglands, now a NNR - then 'a metropolis' for the species. Collier's observations were then encompassed in his valuable book on the conservation of the insect (Collier, 1986). As for the butterfly's demise in England, there is no evidence that overcollecting was responsible, although Collier gives accounts of restrictions imposed by the NCC, including complete collecting bans in all NNRs in the East Midlands, and to two specimens for each licence holder working an NNR. Low populations are always acutely vulnerable to extinction by collectors. Ironically

Flagship species

collection of the Chequered skipper was only prohibited in England when it became extinct, and where it now survives in Scotland it has never been prohibited to collect it.

The conservation strategy formulated by Collier came up with seven points, (i) management policies should be made for all sites that are either SSSIs or NNRs, (ii) collecting without a licence should be illegal, (iii) a working group should be set up (iv) research and monitoring should be carried out, (v) the management of 'wayleaves' under power lines should be actively pursued, (vi) there should be more media awareness of this species, and (vii) reintroduction of the species into England should be considered. I am pleased to report that many of these positive conservation proposals are being implemented.

The Chequered Skipper was put on Part I of the Wildlife and Countryside Act 1981 in September 1982 under Schedule 5 which seeks to protect the species and its breeding sites. This was done 7 years after it became extinct in England. Butterfly Conservation have continued with the impetus of keeping the Chequered Skipper in focus. In 1994 they organised a reconnaissance trip to one of the strongholds of the butterfly in West Scotland (Feltwell, 1994,1995) and in 1995 have embarked upon a plan to re-introduce the butterfly back into England (Butterfly Conservation, Press Release, March 1995).

The Chequered Skipper may have occupied the minds of lepidopterists over the last decade or so as a species worthy of re-establishment, but there are now a number of other species recognised as targets for conservation. Apart from the Species Recovery Programme operated by English Nature which has targeted a few species, action plans are being written for others independently by Butterfly Conservation, as well as by the Biodiversity Challenge Group. At the time of going to press action plans for a few butterflies have been submitted to the Department of the Environment ready for publication, as part of work commissioned from the Biodiversity Challenge Group (see RSPB, 1994 and Wynne *et al.* 1995)

CHAPTER FIVE
Conservation of the Swallowtail

INTRODUCTION

As the largest resident British species, and a spectacular one, the Swallowtail has always drawn lots of attention. Fortunately it frequents very wet areas which are often inaccessible, thus keeping collectors literally at arm's length. (1) It has a straightforward life cycle like most British butterflies, and is associated with one main foodplant in England, but it does, admittedly have occasional problems from parasitic hymenoptera which attack the larvae. However, it does not have the complications of the ant-butterfly relationship typical of the Large Blue, so in view of all these factors it should be easy enough to conserve.

Conservation of the Swallowtail

The Swallowtail has more in common with the Large Copper, at least in England, since the two exploit a particular kind of watery habitat and are therefore amongst the few English butterflies that inhabit wetlands. Also like the Large Copper, the breeding success of the swallowtail is much more dependent on the vagaries of the water table, which have a profound effect on the health of its foodplants.

The Swallowtail that occurs in England is the subspecies *Papilio machaon britannicus* which has a distinct and rich orange background colour compared with the other European subspecies, as well as a slightly different extent of black markings and it is smaller and weaker in flight (Brian Gardiner, pers. comm. 1994). This interesting subspecies does have good grounds for deserving conservation since it is endemic to England. Thirty-two of Britain's butterflies including the Swallowtail have variable colour forms (Dennis, 1993). The conservation point is that the typical English subspecies should be a priority to conserve.

Apart from its former haunts in southern England and the Thames estuary, the Swallowtail currently exists in two separate situations in East Anglia, the Broads (Norfolk) and Wicken Sedge Fen (Cambridgeshire) where it is actively conserved. Being a flagship species, a certain amount of special attention has been afforded it in preference to other notable, though less glamorous insects.

When Wicken Sedge Fen was established at the end of the nineteenth century it became one of the first nature reserves in Britain, if not the world (Watson, 1977). So Wicken is of great historical importance in the world of conservation, and the presence of the Swallowtail helps to make it so.

It was in 1899 that J. C. Moberly, with others bought tracts of fen at Wicken and gave it to the National Trust. The NT run Wicken, and have persistently cleared large areas of alder carr so that typical reedbeds and sedge prosper. Milk-parsley, *Peucedanum palustre,* which is the foodplant of the Swallowtail in Britain, thrives in this typical black fenland soil. It is a local wetland species, and thus especially vulnerable to any drying out of the land. This is a plant and looks rather like Cow-parsley, *Anthriscus sylvestris*, to the inexperienced eye, but it is much sturdier and taller, and which has a very restricted distribution; it is found only in a few isolated parts of England, the largest area being within the fenland and Broads of Norfolk (Fitter, 1977).

CONSERVATION AT WICKEN

There have been two extinctions of the Swallowtail at Wicken, once in the early 1950s (Dempster, 1977) and again, apparently, in 1979 (Duffey & Hall, 1980). When numbers counted fall to single numbers (e.g. only nine butterflies were seen in 1977 and three in 1979) the butterfly has essentially become extinct as it did in 1979. The reason for the extinction in the early 1950s was because the surrounding land had been drained, thus resulting in changes in the habitat at Wicken (Dempster *et al.* 1976). The 1979 extinction was probably due to cool wet weather resulting in few eggs being laid.

The problem with Wicken Sedge Fen is that it is far too small to accommodate such a large butterfly. It is like trying to keep a living collection of butterflies in a cage with open sides. The habitat is not right, it is too dry, the peat has dried out significantly and has become oxidized over the last few decades, and the Swallowtail's foodplant does not grow freely besides the waterways and in the watery meadows as it should. There does not seem to be enough Milk-parsley at Wicken, although there is a special milk-parsley meadow set aside for the butterfly; it is a small concession for such a big insect, which has a wide penchant for freedom. It cannot be expected to stay just there. The Swallowtail is a wanderer turning up in various parts of East Anglia, often having been blown there. It is not a butterfly of small patches and has to be recognized as an energetic wanderer. Thus the small patch of suitable habitat provided at Wicken is a token gesture to this giant of the lepidoptera.

Wicken exists as a specimen 'fen', sadly stuck 3m above the surrounding land, an oasis of semi-natural wetland habitat in a monotony of sterile agricultural land. It is excellent as an educational fen, for an instructional walk around the broadwalks, even to study what thick Alder carr and dead trees are really like, but the chances of seeing a Swallowtail are poor. However, the fact that the Swallowtail might be seen is enough to bring some people in.

The repeated attempts to maintain the butterfly at Wicken are all very laudable, if only to generate a relevant conservation talking point, for the masses of tourists who visit Wicken each year. As a classic wetland site, of international importance in Western Europe, the Swallowtail must still be maintained there despite the site's shortcomings.

How does one view the success of Wicken today? The bill board says it is one of the best examples of a wetland in Europe. Has the conservation of the butterfly been responsible for all this? Certainly it has been the *raison d'être* behind a lot of management in recent decades, and the fen has improved a lot from the dark and dingy Alder carr that covered the site 20 years ago, but its heyday was during the Victorian era when the reeds were cut regularly for the thatching industry. What is certain is that the conservation of this one insect has played its part in the making of a fine wetland habitat

CONSERVATION ECOLOGY

It takes an extinction to focus the conservation mind, so it was in 1971 that Jack Dempster and his colleagues (1976) set about an ecological study of the Swallowtail to discover why its numbers had declined in Norfolk, and declined to extinction at Wicken.

Unfortunately no-one had graced the Swallowtail with extensive and protracted autecological studies, as has befitted some other attractive butterflies such as the Large Blue, Large Copper and the Heath Fritillary. For Swallowtail studies we look to the works of Jack Dempster, Brian Gardiner, Marney Hall, Eric Duffey and a few others. That might be a relief to those who have to plough through reams of the written word, but it does indicate that the Swallowtail is still not as heavily researched as some other species.

There is evidently room for concern for the future of the Swallowtail, since the foodplant is diminishing in abundance, the Broads are becoming increasingly over-loaded by tourists causing problems through their indirect effects, and there are always the unpredictable effects from global warming. History has shown that without extensive autecological research anything could happen to the well-being of the butterfly, and catch all the lepidopterists by surprise. It might be curious to consider that the last stronghold of the Swallowtail is in an artificial habitat, the Broads, and that it is more or less in synchrony with a man-managed environment where reeds are harvested regularly. Most British butterflies, however have colonized artificial habitats. The Broads are a completely different habitat from the one that was there before, and it is this new habitat to which the Swallowtail has made its adaptations. Quite how synchronous was (and is) the interaction between humans and butterfly might never be known, but the butterfly probably had adaptations to survive even when there were seasonal variations in the reed-cutting periods.

One ecological fact unearthed by Dempster and his colleagues, which is of use in conservation management, was that the butterfly lays only on tall, prominent Milk-parsley plants. Wicken tends to produce shorter plants, compared with those on the Broads - an adaptation to the constraints of reed-cutting. It is interesting that the height and vigour of the Swallowtail's foodplant is determined directly by the frequency of harvesting of the reeds, and the cutting of peat, since any human neglect results in overgrowth, plant succession and dominance of Alder carr which smothers all butterfly habitats. Different regimes of reed-cutting have prevailed at Wicken; the traditional one may have been early cutting, whilst now the tendency is to cut in summer and autumn (Rowell, 1987).

Perhaps a bold move for any conservation body such as the National Trust, RSNC or Butterfly Conservation, is to make a temporary commercial leap into the world of peat production to meet the insatiable demand of gardeners, simply by moving into the Broads (and parts of Wicken), removing peat and creating fine butterfly habitat as a result. The Swallowtail would not be in the Broads if it were not for the peat-diggers of the past.

RE-INTRODUCTIONS

Up to 1974, when Eric Duffey published his *Nature Reserves and Wildlife* he mentioned that there had been several attempts to re-introduce the Swallowtail to Wicken but all had been unsuccessful. This was verified by Dempster & Hall (1980) who recalled that Brian Gardiner of Cambridge reared Swallowtails from Norfolk stock and released them in the fen, and that other releases were made by Mr. H. G. Short.

Brian Gardiner had actually bred and released Swallowtails on a regular basis into Wicken. As one of Britain's foremost breeders of butterflies, he embarked on a breeding programme in 1954 on behalf of the Trustees of Wicken Fen to supply mated adults and larvae to restock the Fen (Gardiner, 1960, 1963). From 1957 to 1967 about 50-300 adults were liberated into Wicken by Gardiner, but there had been other people before and after this period who had also re-introduced stock (Brian Gardiner, pers. comm. 1994).

One of these re-introductions involved stock from a Broads population, when in 1972, 50 eggs were taken from Hickling Broads NNR and reared at Monks Wood Experimental Station by Jack Dempster. By the autumn of 1974 over 1,300 pupae were ready for the grand release of butterflies into Wicken in the summer of 1975, to 'inundate the area' but some of these pupae were unfortunately killed by late frosts in June, prior to release. However, all was not lost, and the released butterflies soon laid an estimated 20,000 eggs in the fen, and this probably produced 2000 individuals which made it through to pupation (Dempster & Hall, 1980). However, all this effort resulted in the Swallowtails' population fizzling out to extinction by 1979.

One of Jack Dempster's special interests has been the way in which the morphological characteristics (size, lengths, colours, etc.) of populations of butterflies change according to the locality and through isolation through fragmentation of habitats. However, despite his extensive measurements on museum specimens it was not shown that Swallowtails at Wicken had any 'characteristics which could have accounted for its extinction at Wicken or which seemed likely to prevent its surviving adequately there' (Dempster & Hall, 1980).

Conservation of the Swallowtail at Wicken, at least, is thus seen in the light of new introduced captive-bred stock, not necessarily compatible with the original stock. The butterfly is exceptionally easy to breed at home, and there have been numerous re-introductions in the Broads and at Wicken. All of these will have certainly helped to dilute the Swallowtail's genetic integrity. It was E. B. Ford who said that ...'the Swallow-tail butterfly seems to have survived here, at the edge of its range, by adapting itself closely to a specialised type of habitat in the fens' (Ford, 1945, page 303). There is still something peculiar about the English Swallowtail, its different foodplant compared with the Continental subspecies, its different habitat, and, perhaps different physiological aspects which we have yet to learn; after all, Ford said that 'This butterfly is purely a fen insect in England, though that is not its normal habitat abroad. It is worth noticing that other organisms, for example plants and birds, may be found in abnormal localities near the edge of their range.' The fens and Broads are the furthest areas in western Europe in which the insect occurs, and besides living in a wetland habitat, is also subjected to the damp Atlantic climate. Nothing could be more abnormal or peculiar to Britain. The insect thus should be conserved at all costs, even at Wicken.

One cannot help feeling that attempts at re-introducing the Swallowtail at Wicken might be thwarted by all the adults just blowing away in the first strong wind after release or eclosion. Wicken is flat and is not protected by any tall marginal vegetation (as at Woodwalton Fen), so I was not surprised to note in Ted Ellis's (1965) book on *The Broads* that he had recorded Swallowtails being blown 80 km away after a strong wind. The butterfly had also been recorded as a stray in Northamptonshire (Rothschild, 1975). Clearly when there was much available habitat in the fenland or in the Broads dispersal by wind was beneficial since it helped to found new colonies, stimulate others and generally help in the mixing of genetic material. The fact that nowadays nature reserves exist as tiny fragments in a sea of agriculture bodes very badly for any butterfly found outside its prescribed locality, and does not help conservation activities on these tiny nature reserves. As Ellis (1965) said 'such enforced wandering only serves to deplete the native stock.'

The conservation point about the Swallowtail being a wanderer is that in a wetland habitat (which tends to be large) with plenty of patches of foodplants, it does not really matter if the insects moves around from one patch to another, but in today's world of fragmented parcels of suitable habitat, the casual or determined butterfly wanderer is not going to fare well in the world outside its nicely cosseted reserves. Its populations will always have to be nurtured and topped-up.

CONSERVATION IN THE NORFOLK BROADS

The English Swallowtail has always been a native of the Broads to the east of Norwich since the large peat pits were flooded in late historic times. One who knew the ecology of the Broads was the late Ted Ellis, a Broadsman by adoption, and owner of Wheatfen Broad.

He perceived the status of the Swallowtail as being abundant 'very largely due to the fact that considerable areas of watery fens which used to be mown regularly in summer (and burnt off occasionally in early spring) have been left to run wild during the past thirty years or so.' (Ellis, 1965). In other words between the 1930s and 1960s there was no specific management of the fens for the butterfly and it thrived because, as Ellis says, the chief foodplant of the larvae flourished, and the larvae and pupae were left undisturbed. That was obviously a twilight period of success, since plant succession finally played its part and scrub and trees started to take over the unmanaged fens. That precipitated a decline in the butterfly. At the same time some fens were drained, subjected to herbicides and 'improved' for agriculture.

Ellis (1965) postulated that 'The time may come when, because of local changes in land utilisation, the minimum reserve population of this butterfly necessary for survival will not exist to meet emergencies'. He did not elaborate on the amount of area needed for the butterfly, but he did say that there were extensive nature reserves in the vicinity of the Broads, although the breeding grounds of the Swallowtail formed only a comparatively small proportion of the areas covered by them.

As for collectors he thought that too many of Norfolk's black or dusky swallowtails had been snatched away as prizes for collections, so that a full investigation into this colour morph could not be carried out. Ellis thought strongly that would-be collectors should take a conscious responsibility in safeguarding the Broads for the Swallowtail in its native paradise.

Fortunately one small part of the Broads has been conserved especially for the butterfly, and that is Catfield Fen, mentioned by Ellis as a large swampy area. Butterfly Conservation made a major contribution to Swallowtail conservation by purchasing this fen in for £60,000. Catfield Fen is a 24 ha wetland in the Ant Valley, which has both international status, and is an SSSI (Butterfly Conservation, 1992b). Traditional management regimes and careful control of water levels are being used to maintain the habitat just right for the butterflies in this wetland. (3).

Conservation of the Swallowtail

OVERVIEW OF BOTH POPULATIONS

The Swallowtail is classified as 'Vulnerable' in the *British Red Data Book: 2. Insects* (Shirt, 1987*)*, which means that it could become 'Endangered' (which is the penultimate stage to 'Extinct') if the factors affecting its livelihood continue. The main factor is the continued drainage of adjacent land. According to Jeremy Thomas' account in the above-named book published in 1987, the Swallowtail was successfully breeding in each of the five separate systems within the Broads and was locally abundant in each. It also safely exists within two National Nature Reserves and in at least three Norfolk Naturalists' Trust reserves. The distribution map of the Swallowtail shows only four 10x10 km squares of distribution centred on the Broads and one on Wicken (Heath *et al.* 1984; Emmet & Heath, 1989). The butterfly is confined basically to the rivers Ant, Thurne and Bure, as Emmet & Heath state, and, according to data from Martin Warren there are about 80 colonies of the Swallowtail left (Butterfly Conservation, 1992).

ISOLATION AND GENETIC EFFECTS

Bearing in mind that Ford reminds us that the English population of the Swallowtail is abnormal, and that we should expect this sort of thing, it was no surprise that Dempster found slight differences between Norfolk and Wicken specimens. Those from Wicken had slightly longer wings (Dempster et al 1976). This was extrapolated by Heath *et al.* (1984) to indicate that smaller wing size might restrict its powers of dispersal. This might actually be a good thing helping to restrict the butterfly to its carefully measured nature reserve, rather than a bad trait of genetic isolation.

Dempster *et al.* (1976) indicated that specimens of the British subspecies are nearer to Maltese and other Mediterranean subspecies than to others, even nearer ones across the Channel, which says more perhaps about isolation on islands than about evolution on large land masses. Isolation can cause size and colour to change, sometimes significantly from mainland specimens. If Wicken became progressively isolated up to the 1950s then differences from the Norfolk Swallowtails might be predicted. Since, however, they have fairly similar characteristics, Dempster *et al.* (1976) suggested that no evolutionary changes had occurred that would affect any chances of establishing Norfolk butterflies at Wicken. The justification for re-introductions was therefore set.

THREATS

As explained earlier, the Swallowtail is affected not so much by direct impact from humans as by habitat loss, i.e. loss of fenland whose wet meadows with Milk-parsley disappear through neglect and plant succession. Wet meadows that are never grazed or cut dry out and develop a scrub of various water-loving species, such as Sallows and Willow (*Salix* spp.) and Alder (*Alnus glutinosa*). This is made worse, and the rate quickened, by the general draining of surrounding land which goes on all the time, and has especially been the case of farmed fenland in which some of these butterfly habitats occur. In the Broads increasing tourism results in degradation of waterside banks through erosion from boat back-washes, and thus in the degradation of Milk-parsley populations. The effect on butterfly foodplants has a knock-on effect on the butterfly population.

Few butterflies are likely to be affected negatively by an increase in sea level if global warming should occur in Britain, but the Swallowtail is one of them (Dennis, 1993). The scenario, one imagines, is that the sea water will percolate up the myriad of channels from the source of the River Yar which drains the Broads at Great Yarmouth, and put many of the Milk-parsley plants under saltwater. This would be a further blow to the butterfly. A slow rise in sea level would be more helpful to the butterfly than a quick one, since the butterfly would have time to exploit suitable pastures, if there were any left in the agricultural hinterland.

In summary, the Swallowtail exists in two places in Britain, the Norfolk Broads and Wicken. It seems to have perpetuated itself in the Broads since they were naturally flooded, because the Broads are such a large ecosystem of interconnected habitat, some of which is suitable for the butterfly. Wicken is just a small fragmented fen stuck in the middle of agriculturally 'improved' land with no inter-connections to any other suitable habitat. The academic community regard much of the work of introducing the Swallowtail at Wicken as a failure (Morris, 1986 quoting Dempster & Hall, 1980 as 'conservation had not proved possible'). I am rather the optimist who likes to accept the experiments at Wicken as a success since there is always a chance of seeing this gem of the wetlands, provided sufficient numbers of butterflies have been introduced in any one year. The fact that the butterfly has not established itself sustainably is regarded as the failure. In the Norfolk Broads its future seems reasonably secure for the moment, and, barring an unforeseen major environmental impact, it should continue to grace these particular wetlands of Britain.

Conservation of the Swallowtail

Notes:

1. Not only are the fens and broads especially wet, but various levels and wetlands of the Swallowtail's former haunts were also relatively inaccessible, such as the 'Pulborough Levels' of West Sussex where Stainton (1857) recorded it as common, and, citing Bretherton (1951a) the marshes on the Rivers Thames and Lea (Dempster et al. (1976).

2. For full details readers are recommended to browse through the journal 'Nature in Cambridgeshire' which gives annual sightings and tit-bits of information.

3. T. Douglas Fearnehough (1938) recalls a hunting trip 'From here (Weyford Bridge) I rowed downstream to the junction of the River Ant and Stalham Dyke, in which district the Swallowtail has its haunts....One batch of Swallowtails, (easily recognised for one of them was a dark ochreous variety) passed three times....each time I missed!...The total "bag" for the day was two females, five males (all perfect) half a dozen eggs and about twenty young larvae...' This is close to British Conservation's Catfield Fen SSSI.

CHAPTER SIX
Conservation of the Large Copper

'It has always been a matter of the keenest regret, not only to Lepidopterists, but to all Nature lovers, that we should have lost our peerless native race, for <u>dispar</u> is unquestionably the most beautiful of all European butterflies when seen in its natural surroundings, and, of all the races, the old British one was the finest. It is very beautiful as a cabinet specimen, but it must be seen in flight for its full beauty to be adequately appreciated.'   Anonymous, 1929, page 57.

BACKGROUND TO ITS CONSERVATION

This account of the Large Copper introduction is drawn from a variety of sources, but most important of all has been Anonymous (1929 - 'almost certainly Lord Rothschild and Karl Jordan', Miriam Rothschild pers. comm. 1994) backed up with Duffey's (1968) comprehensive account. There was a casual introduction of

Conservation of the Large Copper

continental Large Coppers to Wicken Fen by George Verrall and James Tutt in the early 1900s and there have been various discussions on the virtues of Large Copper introductions, including those of Bink (1970) and Van Schepdael (1962). More recently Andrew Pullin and Mark Webb at Keele University have renewed Large Copper conservation in Britain (Pullin *et al.* 1993, 1995), and there has been a gradual drift away from talking about introductions from the Continent as 'introductions' to one of 're-establishment' in recognition of the similarity of the subspecies.

The first British Large Copper was described in 1749 when it could be found in the fenland areas of Cambridgeshire, Norfolk and Suffolk. It was pursued vigorously by collectors for the next few decades since it was regarded as one of the jewels of the British butterflies - 'the glory of British entomology'. Collectors were almost entirely responsible for the butterfly's extinction and the last known specimens were collected by Stretton at Holme Fen in 1847/8 (Anonymous, 1929) and at 'Whittesley Mere' (the current spelling is Whittlesey) in 1851 after drainage and reclamation at one of its traditional haunts (Duffey, 1974) or at Quy Fen or Bottisham (Cambridgeshire) in 1851 (Dennis, 1977). However, Irwin (1984) reported a collector with two Large Coppers with labels stating Ranworth (Norfolk) 1860 and Woodbastwick 1864 which gives some evidence of the Large Copper surviving later in the Broads. Norman Riley argued, quite plausibly, that drainage was not entirely responsible for extinction at Whittlesey Mere since drainage occurred 5-6 years after the last capture of the Large Copper; his resume of the copper collectors in the fens leaves no doubt about the butterfly's decline. The circumstances for the demise of the Large Copper were reviewed extensively by Pullin *et al.* (1993, 1995), however, who demonstrated that the extensive drainage in the fenland around The Wash during the seventeenth and eighteenth centuries effectively eliminated populations through outright habitat loss.

It is probable that a combination of over-collecting and drainage was responsible. The jewel had lasted about 100 years from discovery to extinction. The Large Copper had been in the region for probably 10,000 years at least, but it had been extinguished in a tiny fraction of that time.

The British population always had a single (univoltine) generation each year, unlike some bivoltine subspecies on the Continent such as *Lycaena dispar carueli*, the commonest subspecies in France. Univoltine subspecies on the Continent include *Lycaena dispar batavus* in Holland, and *Lycaena dispar gronieri* in France (St. Quentin). It was *Lycaena dispar batavus* that was used for introduction to the English fenlands. (1)

RE-INTRODUCTIONS

1909 re-introductions
There is a gap in the history of the conservation of the Large Copper of about 60 years between when it became extinct and its first re-introduction. The first place chosen was Wicken Fen, since it was there that G. H. Verrall 'turned down' a number of larvae onto Water Dock, *Rumex hydrolapathum* in 1909 (Verrall, 1909). He released specimens of the continental subspecies called *rutilus,* but few details of this small and modest operation are documented (Duffey, 1968).

Verrall was a little apologetic about the re-introduction that he and his colleague Mr. J. W. Tutt arranged at Wicken, excusing themselves on the grounds that the last British specimen was caught in 1848. He called on people not to collect these Wicken coppers, or their progeny and hoped that these butterflies would revert to the British form. According to Anonymous (1929) this introduction was not successful. Wicken had not been a classic site for the Large Copper (although there is a specimen in The Natural History Museum which is from Wicken Fen in 1830) and in the early 1900s Wicken was presumably still quite a viable wetland site, not so raised above the surrounding countryside as it is now. There was to be another release at Wicken in 1930, but overall Wicken has never figured strongly in Large Copper conservation.

1913 re-introductions
The second attempt was in 1913 in a snipe bog at Greenfields, Tipperary in Eire, again using *rutilus* stock. Specimens for release were obtained from the Heningsdorf and Finkenkrug marshes north of Berlin (Germany). Prior to the first release the Eire habitat was managed by clearing scrub, importing and establishing English Water Dock roots and planting seedlings. Ten dozen larvae of *rutilus* were sent from Berlin by Herr H. Rangnow but most died after they were transferred onto potted plants. Only eight adults hatched and of these the four females mated and laid in the wild.

In the following year (1914) Captain E. Bagwell Purefoy travelled to Berlin, met with Herr Rangnow and his son, collected nearly 700 larvae brought them back to Eire; and from these some 400 adults emerged for release on the site after sustaining quite a high level of parasitism. It is interesting that efforts were made to stop any German parasites being released into Eire by restricting the breeding stock to nets. This final stocking was a great success, and resulted in the colony in the snipe bog becoming firmly established. It was reported to be in a healthy condition 14 years later (1928) (Anonymous, 1929). Thereafter it appears to have declined and is not reported again.

Conservation of the Large Copper

1916 re-introductions
The success of the Irish experiment drew the attention of Hon. N. C. Rothschild of Tring who asked Captain Purefoy to check out the habitat at Wood Walton Fen (as it was known then) in 1916 prior to a possible introduction of stock from Eire. In about 1916 about 20 acres (8.1 ha) of scrub (stunted alder (*Alnus glutinosa*), birch (*Betula* sp.) sallow (*Salix* sp.) and aspen (*Populus tremula*) was cleared to prepare the butterfly habitat.

Despite the English Large Copper butterfly having become extinct in 1847/8, Rothschild had exhibited specimens of Large Coppers from the Netherlands (obtained by Mr. R .A. Polak) at the Entomological Society of London at which he stated they 'were identical with the extinct British race' (Rothschild, 1915a); Hungarian Large Coppers were also exhibited at the same time. The specimens from the Netherlands were very similar, especially on the uppersides, but later examination of other stock indicated to others that there were perceptible enough differences to warrant subspecies status. Thus the Friesland specimens were named *batavus* by Charles Oberthür in 1923. The differences in spotting only could be seen when comparing long series of butterflies (20-50 at a time). The English subspecies had been called *dispar*, and was regarded as the most attractive of all Large Copper subspecies in Europe. (2)

Conservation of Large Copper populations on the Continent has always figured strongly in the minds of those who conserve the Large Copper in Britain. The urgency to establish the Large Copper in England (albeit the *batavus* subspecies) was heightened by the drainage of one copper habitat in Friesland in 1915 and threats from collectors at another. It was thought likely that the Large Copper could become extinct in the Netherlands too, thus denying a resource to replenish the English stock.

1926 re-introductions
The next initiative to establish *dispar* in England came via the Standing Committee for the Protection of British Insects established by the Royal Entomological Society of London (RESL). Captain Purefoy found a suitable site on private ground at Woodbastwick Hall in the Broads (Norfolk) and 550 *rutilus* pupae from Eire were brought over.

The establishment was a success so the Committee looked around for a suitable nature reserve where the insect could be maintained. This is where Woodwalton came in, since the custodians, the SPNR, offered the fen to the Committee as a suitable place for the butterfly. On his death in 1923, Charles Rothschild bequeathed the site to the SPNR, and as it was the presumed intention of Rothschild to establish the Large Copper there the SPNR took on his aspirations. Woodwalton lies 'in the heart of the dispar country' not far from Holme Fen where some of the last English specimens were netted.

The first task of the Committee was to have the area which Rothschild had cleared in 1916 cleared again of all scrub, and to prepare the 8.9 ha site for the butterfly. Lord Rothschild paid for the cost of the clearance. Attendance to the precise oviposition sites of the female was paramount in the conservators minds. Talking of the Water Dock, Anonymous, 1929 said:...'it is extremely difficult to get it to grow on the flat, and surrounded by natural cover of a height to suit the butterfly; it will apparently only do this when the soil is peat, and the terrain a very low-lying one, with the water never very far from the surface. Such ground must inevitably be subject to floods, in fact it always is, but it is on plants growing under such conditions that the female butterfly prefers to deposit her ova.'

During 1926 (and later in spring 1927) innumerable clumps of Water Docks were planted through the area to receive the butterflies, and the annual cutting of hay was done during the appropriate period of May. This allowed the site to become an ideal 3-4 feet deep in vegetation during the breeding season.

The Committee needed to get specimens from the dwindling Holland populations so they liaised with Mr. A. J. Labouchere of Utrecht and Mr. J. H. E. Wittpen of Amsterdam, the latter being an expert in breeding *batavus* in captivity. The summer of 1926 was terrible and Wittpen succeeded only in capturing two females which resulted in 'a very small number of eggs' for Purefoy at Woodwalton.

1927 re-introductions
In the following year (1927) a posse of lepidopterists including Captain Purefoy, James Schofield, Labouchere and Wittpen went to the *batavus* marsh in Holland to collect stock. They found only five females and a few larvae, which collectively produced for Woodwalton a total of 38 specimens of which 13 were females. The adults were released into the fen, and later *in coitus* pairs were found and collected (not an inconsiderable feat for the fen watcher) and these confined to potted Water Docks for a few days to lay eggs. The total egg count thereafter was 300-400. During August, Wittpen secured six more females 'somewhere in Friesland' and these resulted in several hundred larvae reaching the hibernating stage.

Woodwalton was then hit by severe storms in late December and early January, combined with severe flooding. The average depth of flood water was exceeded, and the whole fen was underwater during a continuous flood of 60 days from 22 December 1927 to 20 February 1928. Purefoy searched the site in March and was surprised to find hibernating larvae which had apparently survived 60 days totally underwater. (3)

The population of Large Coppers at Woodwalton had been firmly established by the summer of 1928, when several hundred males were seen on the wing, followed by an emergence of females. Over 1000 butterflies were further released, matings observed, and ova seen on foodplants. On 15 September Captain Purefoy reckoned that the future success of this experimental colony seemed assured.

1930 re-introduction

Wicken was the target release point for more *batavus* (source not indicated) in 1930, and it apparently survived here for 12 years (until 1942) when Adventurer's Fen (adjacent to Wicken) was reclaimed for agriculture (Duffey, 1974) however, it may have been hanging on here a little longer, since in the NHM there is a specimen dated 2 July 1947 attributed to none other than E. B. Purefoy (amongst others dated 1927 and 1939 for Wicken).

During 1927-1955 there is little documentary evidence of what really went on with regards to maintaining the populations of Large Coppers in East Anglia. Duffey (1968) states that Purefoy remained active during this time advising the warden of Woodwalton on how to look after the stock. There was a re-introduction at Wheatfen Broad in 1949 and this is recounted in Ellis (1965).

CAPTIVE BREEDING

From at least 1930 onwards that the habit of collecting material from the wild and looking after it in captivity became a regular event. The species, it would seem, is now entirely dependent upon survival in this semi-captive state. The butterfly had previously occurred all over the fenland wetlands, unaided by humans, because there were a sufficiently large number of habitats available. Now that the species is restricted to habitats that are too small to maintain viable populations it has to be bred in captivity. As Duffey (1968) summarizes: 'it appears unlikely that a viable population can be permanently maintained under natural conditions because the habitat is too restricted and the population will probably never be large enough to withstand fluctuations in numbers caused by predation, parasitism and adverse climatic conditions.' It was also during this period that the famous 'Copper Fields' were prepared in Woodwalton Fen: an area maintained today, where visitors may get their chance of seeing Large Coppers. When Woodwalton Fen was declared a National Nature Reserve in 1953/4 the Large Copper population was confined to two areas totalling 1.8 ha.

1960 MANAGEMENT

Active conservation began in 1960 with initiatives from Monks Wood Experimental Station (Monks Wood Experimental Station, Annual Reports from

1960). The old custom of taking stock from the wild involved collecting young larvae from their foodplants in the autumn and looking after them in captivity safe from predators and parasites. This was to secure a larger number of adults than would otherwise be achieved through the usual gambit of natural wastage. It had been established that up to 95% of the early stages suffered parasitic attack thus dictating this conservation practice. There were a series of introductions of the Large Copper from 1960 onwards co-ordinated by Eric Duffey.

Problems with parasitism and predation caused the species to have a precarious existence at Woodwalton, and its presence there was only due to continued support from humans during the larval stage. Extinction was always a possibility, indeed there were two occasions when the species might have become extinct between 1960-64 though parasitism by the fly *Phryx vulgaris*. Duffey subsequently found that predation from mammals and birds was six times more significant than the effects of parasitism and predation from invertebrates; this he established from captive breeding under nets.

For the 1960 introductions a 6 ha part of Woodwalton was prepared and planted with 450 Water Dock plants, and to this habitat were released 11 males and 13 females. This resulted in a 1961 count of 20-25 adults and 1,103 eggs and newly hatched larvae. By the following year only seven to ten butterflies were counted; the reduction in population was blamed on predation and parasitism. 1963 was a better year with 2,117 eggs and larvae counted in the wild, and 9,567 around the experimental area. 1964 was even better.

As Captain Purefoy had found earlier at Woodwalton the main problem for conservation was an integrated ecological one: getting the foodplant right. The irony is that the Water Dock germinates in a film of water, or in water, right on the edge of waterways, but it is precisely these plants that the female apparently avoided when laying eggs, preferring plants growing in the fields. To meet the ecological demands of the butterfly, Duffey devised a system of shallow scrapes, so that flood water would carry seeds to the scrapes where they would germinate. They did, and provided in due course highly suitable oviposition sites for the butterflies. (4)

1965-1969 MANAGEMENT

To combat the continual problem of encroachment from vegetation, some grazing had been taking place at Woodwalton since 1957, and in 1965 33 Friesian bullocks were put on part of the fen from January to mid-March. In conservation terms, cattle are highly beneficial since they help to break up litter, encourage grasses and heather and generally clear vegetation from the previous year. This is good for butterfly populations. Some of the older breeds of cattle are more useful

Conservation of the Large Copper

for clearing purposes than more recent ones, and in January 1966 ten Galloway yearling steers were put on the south-eastern part of the fen.

In contrast recent experiments in the Bordeaux area of France with the re-establishment of *L. dispar bordielensis* it has been found that grazing with horses is much better than with cattle. The result of experiments comparing the two types of grazing in a suitable habitat, with Curled Dock, *Rumex crispus* (thought better than *R. hydolapathum*) will be published shortly (Lhonoré pers. comm. 1994).

During the late 1960s there was some confusion regarding the way Large Copper conservation was going. Duffey (1968) stated that Woodwalton was too small to sustain a viable population of the butterflies, whilst at the same time the NC (1968) initiated further research into the autecology of the insect: 'The project was primarily concerned with finding out whether it was possible to establish a permanent natural population of the Large Copper on the area of fen where suitable habitat conditions could be created. At present there is about 6 ha of breeding ground (out of a total of 207) and it is thought that the maximum might reach about 60 ha. This would take many years to achieve.'

In retrospect the early work and successes of reinstating the Large Copper at Woodwalton had never been viable in the long term. The NC made the decision in 1968 to instigate further work on the Large Copper, but the NC, and later the NCC, failed to follow it up, or to seek alternative habitats, because *L. d. batavus* was considered to be an alien species and therefore of lower priority than the threatened native species (Sheppard, pers. comm. 1994).

1969 EXTINCTION OF RE-INTRODUCED STOCK AND SUBSEQUENT RE-INTRODUCTION

The Woodwalton Large Copper *batavus* population died out in 1969, having had a reasonably clear run since 1927/8 (Monks Wood, 1969-71). The female Large Copper could not find the foodplants on which to oviposit, not because of extra plant growth obscuring them, but because of summer flood water submerging the plants. Fortunately some Woodwalton Large Copper *batavus* stock had been held elsewhere and some of this was prepared for re-introduction back to Woodwalton. In 1970, Mr. H.G. Short provided the NC with 1,273 larvae from his home in Esher (Surrey) and these were reared on Water Dock at the fen and protected under muslin to minimize the effects of predators and parasites. The total number of butterflies released at Woodwalton that year (including 52 males and 167 females bred at Monks Wood) was 517 males and 551 females which produced 2,672 eggs, or 4.8 eggs per female. These releases produced 344 males and 208 females in 1971 which was a warm, dry summer and they gave an autumn egg count of 6,509, or 31.2 eggs per female - an improvement. A female can lay over 300 eggs in semi-captivity.

CONSERVATION CRITERIA

Successful protection and conservation of butterflies is dependent on a thorough knowledge of their biology and habitat requirements. With this ideal in mind, and before the establishment of the NC, the energetic work of Lord Rothschild and the CPBL did a lot to further the conservation objectives as we would perceive them today. That the early attempts to introduce the Large Copper to Woodwalton were so successful was due entirely to the very best autecological studies that could have been done at the time. However those introductions were too small and did not result in sustainable viable populations. If it were not for the energies of Lord Rothschild, who had a particular fondness for butterflies, this early work on the conservation of the Large Copper would probably never have happened. There was some considerable feeling at that time that the Large Copper introduced at Woodwalton was an alien species, and as such, there was not so much enthusiasm for continued research.

If the goal of continued research in conservation of a species, then it is important to get out into the field and see what is going on in its habitat - to understand more about inter-relationships between the insect and the environment.

Conservationists can budget for 95% wastage of butterfly populations from parasites and predators at specific points in the life cycle and make appropriate conservation management plans, but detailed information on the effect and timing of attacks by parasites and predators is not likely to be of fundamental use in conservation management. General principles will suffice. In the Large Copper the points in the life cycle most vulnerable to predators and parasites are fairly well assessed already, and the culprits pretty well pinned down.

The point is that the criteria for conservation must regularly be examined and questioned. Is perhaps the habitat more important fluctuations in the numbers of parasites and predators over which humans have very little control? Conservationists can control and regulate the structure of the habitat and its floristic make-up much more easily. Is perhaps this a more important and key area to attend to?

One has the impression that some butterfly research is carried out for its own sake, rather than for the conservation of the particular species. There is also much repetition of ecological field work, sometimes done in ignorance of previous work. Overall one cannot but be amazed at how much meticulous ecological work was carried out earlier this century, as expounded upon in

Conservation of the Large Copper

Anonymous (1929); this great body of work contrasts markedly with the snippets of information on the Large Copper to be read in the annual reports from Monks Wood.

CONSERVATION OF PREDATORS AND PARASITES

Predators and parasites have an equal right to be conserved as their prey and hosts; but many of them have missed the limelight as indicator or keystone species and their conservation has often been neglected. However, one now perceives a greater understanding and consideration towards predators and parasites than before, and it is interesting that an awareness of natural enemies is creeping into provisions, codes of practice and legislation. Gone are the days when predators and parasites were deliberately excluded, now they are part of the ecological scene, a complimentary part which may help to regulate populations and should be equally conserved. Some committed lepidopterists are known to wax lyrical about their host parasites if given half a chance, even to find them more interesting than their hosts.

In terms of how to conserve butterflies, the knowledge of which predator and which parasite is causing the damage is academic. Which stage it hits and to what effect are far more important, and, together with the frequency of the attack are the factors with which conservationists have to live, rather than take any remedial action against. For predators such as the Whitethroat, *Sylvia communis* as an example (quoted from Ellis in Duffey, 1968), it is impossible to exercise any control over hungry migrant birds descending to feed on adult Large Coppers, but at least bird boxes should be banned from all butterfly reserves.

That scientists at Monks Wood changed their views on the natural enemies of the Large Coppers (Monks Wood, 1966-1968, page 39) is largely immaterial to the way that the conservation is carried out, since it is currently impossible to control predators and parasites in the wild. The keenness to quantify the natural enemies of the Large Copper in the field had led Monks Wood scientists to reflect differently on which parasites or predators were causing the impact on populations. It was formerly thought that tachinid flies and hymenopterous parasites caused the most damage, but in 1969 it was shown that over 70% of summer larval mortality was due to predation from birds.

High levels of parasitism are an expected external factor in any ecological study of the Large Copper (and many other butterflies too), and had been known for decades; it comes with the ecological territory, and one just has to be wary of it. There is, however, a need to be wary of natural enemies when embarking on a course of captive-breeding.

If one sets a wider conservation perspective in the realms of invertebrates generally, then one has to accept that predators and parasites should be conserved as much as their hosts. Inadvertently in the past, although money for research has not been previously poured in their direction, parasites have prospered and been well dispersed with the comings and goings of Large Copper stock in Western Europe. Historically the objective for conservation has been to conserve the butterfly, since none of the parasites or predators can ever claim to have had keystone status. In more recent years however, the objectives have included the well-being of the habitat and all its incumbents.

FUTURE CONSERVATION

Woodwalton is too small to support a viable population of the Large Copper. Duffey and Pullin both agree on this point. If advocates of butterfly conservation and statutory guardians of the countryside push through their controversial, and aggressive plans, some of which are single-mindedly in favour of butterflies (but which may not receive universal or government consent), we could see some of the smaller nature reserves joined-up, for instance Woodwalton with Holme Fen.

The criteria for butterfly nature conservation must continue to be questioned. Will it be sufficient to have cosmetic nature reserves topped up with some symbolistic / keystone / indicator insects nurtured under plastic, or a viable nature reserve which produces the results using a 'hands-off' approach? Neither Woodwalton Fen nor Wicken Fen is viable since both are raised above the surrounding agricultural land by about 4 m because of drainage and drying out of the surrounding land, and Woodwalton is under the control of the Middle Level Commission which uses the reserve as an emergency site to dump excess water which surges along the large open fenland drains.

With the recognition that the English and the Netherlands Large Coppers ones are the same, Andrew Pullin and Mark Webb are able to grasp the European nettle and think in terms of much wider conservation considerations. The Friesland populations were always treated with some special concern and protection in the 1920s since it was thought that they could become extinct before specimens could be acquired for introduction into Woodwalton. Today Large Coppers are still extant in the Netherlands wetlands and there is close liaison between the nature conservation bodies and researchers in both countries to maintain populations. As the ecological requirements of the butterfly continue to be unravelled, especially with the current works of Mark Webb, the conservation of the species could not be in better hands.

The latest initiative on the Large Copper involves not only Butterfly Conservation but also English Nature. It is one of a few British insects targeted

Conservation of the Large Copper

by EN for special conservation and protection through its well publicized Species Recovery Programme. Butterfly Conservation have now taken on the responsibility for monitoring the population of Large Coppers at Woodwalton. The sorts of considerations that have to be addressed and incorporated into a management plan for this butterfly are (i) open sunny areas for the males to engage in territorial behaviour, (ii) larval foodplants growing in the right habitat, (iii) lots of nectar-feeding stations available during egg-laying and (iv) lots of inter-locking habitats for the butterflies providing an overall big habitat in which the butterflies can survive.

LARGER HABITATS

The Large Copper is a highly mobile species with specimens recorded up to 20 km away from suitable habitat. It needs space, which only huge wetlands can give. Protagonists of metapopulation theory might find it impossible to quantify how populations of Large coppers could survive in a wetland ecosystem given the great mobility of the insect. Antagonists might disagree entirely with metapopulation theory in that there is never such a thing as a closed population since a small percentage of individuals exchange populations.

There is only a little evidence of Large Coppers in the Broads of Norfolk, and, as Irwin quite rightly says, the focus of attention in the mid-nineteenth century was the improved fenland which was much more accessible. But because of their inaccessible nature, the Broads as a habitat may have been considerably overlooked for the Large Copper. The idea of reinstating the butterfly there is highly plausible since it is a large wetland system, According to Duffey (1968) it looks likely that the butterfly had always been in the Broads area and then extended its range into the fenlands about 4,500 years ago when peat started to be laid down (this is gathered from the results of pollen analysis - Duffey, 1968). For a species that is really at the north-west of its European range its distribution has been regulated by changes in habitat caused by the last glaciation period 6,500-7,000 years ago when Britain and the Netherlands were separated.

The Large Copper, as it stands in Britain at the present time, is in a peculiar situation since it is assumed that it can never be self-sustaining in its present locations. However, this may all change if the Broads are managed for the butterflies. Any future habitats need to be big, really big with a huge grid of interlocking waterways and pools. It is salutary to reflect that there were about 2240 ha of continuous fenland available to the butterfly in 1824 in the area of Woodwalton and Whittlesey Mere (Duffey, 1968). Woodwalton will remain an historical but tiny emporium for the butterfly, a tiny pinprick in what was once a huge fenland ecosystem. The highly mobile Large Copper will not fit tidily into such neat little populations. The Broads is altogether a much better place for the long-term safety of the butterfly with its complex of river valleys. We look forward to its careful conservation management there too. Releases at Woodwalton have been big, but have not always resulted in long-term sustainable populations: 700 individuals in the 1920s, and 1,068 in 1969. Some fundamental ecological principles come into play here. Some argue that numbers are not important, since any habitat will only take the numbers of that insect that it can accommodate. If this 'carrying capacity' is exceeded it follows that the extras will be superfluous. In other words it will be wasting stock to add more than is necessary; it may also reduce the chances of success because of lack of oviposition sites. However, until precise figures are given for the carrying capacity of each habitat for each species, we do not have the ecological evidence on which to base a proper judgement. Until that time I am of the opinion that the more that are released the better, and that the fewer will diminish chances of establishment.

CONSERVATION GAINS

As in the conservation of the Swallowtail at Wicken one has to say that the endeavours to preserve the Large Copper at Woodwalton have succeeded in terms of the conservation of the whole habitat (see Duffey & Mason 1970). This has to be applauded, athough it must be said that at no time has there been a positive strategy to conserve the whole habitat.

Ultimately the diligence of naturalists has won against the wholesale destruction of the wetlands, so that conservators are left with pinprick sites in a sea of 'improved' agricultural land. And muddling through these troubled waters of habitat destruction, lack of finance, re-organization and, questionable criteria, these tiny fragments which we call nature reserves have survived (albeit arguably too small and not viable ones for long term solutions - but ideal for eco-touristic and educational solutions) all on the back of one of the two most impressive of British insects. There can be few British insects which have served the conservation movement so well as the Large Copper and the Swallowtail, which will remain symbols of the once extensive wetlands of Britain.

Conservation of the Large Copper

Notes

1. I have examined all the Large Coppers held in the Natural History Museum and conclude that there is no discernible difference between males of dispar, rutilus or batavus, but females of dispar are darker than those of rutilus or batavus; and it seems that Irish rutilus (both sexes) are smaller than either dispar or batavus. Of British material, I inspected 97 females and 119 males of the extinct L. dispar dispar, alongside 17 Irish rutilus and a tray of 64 L. d. batavus. It is noticeable that variation amongst males of Large Coppers is much less in males than in females. It is worth drawing attention here to the published work by Gerard Bernardi on the question of L. dispar and L. d. gronieri, although it lies a little outside the scope of this book: Bernardi, G. (1963). La Rehabilitation du Lycaena dispar dit de Saint-Quentin. Alexanor 3:(2)51-59, 3:(4)9-16.

2. Subspecies or not? Controversy has raged ever since on the status of the two subspecies, L. d. dispar and L. d. batavus. Nathaniel Rothschild and Hemming thought they were similar, Norman Riley and J. Th. Oudemans discussed them as races and Van Schepdael (1962) believed they were the same - just relics cut off by the rising waters after the last Ice Age. Dennis (1977) refers to batavus as a 'close relative' and 'substitute'. Today, after years of debate, current research workers on Large Copper conservation still believe that the two are one and the same.

From his studies, Andrew Pullin is clear in his mind that the two reputed subspecies on either side of the English Channel are the same; that there is no genetic barrier between the two populations, and that there is only weak evidence for any sort of distinction (Pullin et al. 1993, 1995). The fact that the populations were cut off from each other about 8000 years ago so that they could not freely intermix their genes (gene-flow) has done little to separate these populations on their bright external features.

Perhaps the subspecies debate can now be laid to rest. Many people agree that there is little difference at all between dispar and batavus although it remains to be proven that they are genotypically the same; perhaps with DNA studies, using an old technique but relatively newly applied to the world of butterflies, new discoveries will be made. It will throw the whole debate into disarray if significant differences are shown.

3. So interested was Captain Purefoy in the capabilities of submerged larvae that he set about a series of experiments which eventually demonstrated conclusively that larvae can live perfectly well underwater without having to come up for air. In some of his water-filled pots he observed small larvae clinging to leaves for up to 15 days, one even entering the water from a protruding leaf, all apparently oblivious to their watery environment: '..three more larvae became visible on grasses far below the water line.' 'They clung tightly to the stems evidently feeling that their safety depended on their hold.' The water was drained off, and the larvae inspected under a microscope: 'In about ten minutes' time a slight movement was observed, and then the little tortoise-like head was thrust out and the larvae looked around in alarm. It was evidently waking up from a complete trance; vitality was restored to it very gradually from the head downwards, nearly three-quarters of an hour elapsing before it could move its posterior segment. Finally, this larva walked about as if nothing had happened to it.' The average bad flood at Woodwalton does not last longer than 10-12 days, but hibernating

Conservation of the Large Copper

larvae have been known to survive 60 days completely submerged. This great ecological feat is fascinating, not least for how conservation plans can take this into consideration. It is thought that the larvae can survive only this ordeal if they are in diapause which puts them in an appropriate physiological state. With typical low respiration rate and low utilization of energy resources the larvae have good chances of survival in their suspended animation. Other advantages are that their spiracles are too small for water to penetrate and drown them, and to reduce any likelihood of penetration through its spiracles at the front of its body, the larvae have a telescopic head. The spiracles are probably not smaller than any other butterfly species' larvae (though this has not been checked out as far as is known) but the fact that it can retract its first few segments means that it would reduce chances of water entry simply by making less surface area containing spiracles available for water contact, or by covering them in folds of cuticle. There must also be some physiological means by which diapause larvae can sustain immersion whereas actively feeding larvae (i.e. not in diapause) apparently drown if there is an early winter flood. This needs to be investigated too. It would appear from Purefoy's experiments that the larvae choose old brown leaves low down in the fenland debris of leaves and mosses, on which to hibernate. Another conservation consideration is that conservators have to manage, in Britain at least, for flooding at different times in the life cycle, and that flooding earlier in the year is likely to be more damaging than later. If flooding occurs early in the autumn this will delay or prevent the downward movement of the larvae from the Water Dock stems to their hibernation quarters in the litter and peat.

4. It was the intention of Dr Jeffery Harrison to create a similar Large Copper habitat at Sevenoaks, Kent, around the gravel pit nature reserve (Bradbourne Lakes) he created there. I gave advice in the mid 1970s as to the disposition of the scrapes which were duly created using voluntary conservation help from local schools. Unfortunately the untimely death of Dr Harrison failed to realise the fruits of these labours and no butterfly establishments were made.

CHAPTER SEVEN
Conservation of the Large Blue

'The story has a moral....It is a case study in the extinction of a species.' Derek Ratcliffe, 1979.

'As things turned out myxomatosis did not result in the extinction of any plant or animal, with the probable exception of the Large Blue Butterfly' Moore, 1987 p. 141. Norman Moore was Chief Advisory Officer of the NCC when the butterfly became extinct in 1979

inevitable?....'a stab at the heart of Darwinism' Vane-Wright, 1977

'Conservation biology is too full of attempts to rescue species or ecosystems which are transients in the dynamic procession of life...' Robert J. Berry, 1992

'killed by kindness' Jeremy Thomas, 1989a

Conservation of the Large Blue

INTRODUCTION

The Large Blue has been at the centre of butterfly nature conservation in Britain for a long time. It has attracted more effort and resources than most other butterflies, so it is useful to analyse here how the Large Blue became so favoured, and whether there are any lessons learnt for the conservation of butterflies generally.

This chapter starts earlier in the twentieth century when the influential Lord Rothschild became involved with the preservation of species and habitats. At the focal point of butterfly conservation, then, was The Standing Committee for the Protection of British Insects (SCPBI), described in Chapter 2. The Large Blue figured strongly in their endeavours, at least, to start with.

The Large Blue needs few words of introduction. It is a flagship species which has ventured into the public domain through its plight. However, few people have ever seen this species, at least in Britain. The fact that it became officially extinct in 1979 (NCC, 1981) **(1)** means that no-one will ever see the English subspecies of the Large Blue, *Maculinea arion arion* again, alive; there are museums to see them in matched against their evocative localities, a memory of rural England. The Natural History Museum in South Kensington, London, is said to have 1,200 specimens (Spooner, 1963); but actually has far more.

The history of the conservation of the Large Blue butterfly is rather an embarrassment to the statutory authorities and the other private body involved with its conservation. Both the Nature Conservancy (NC) and the Nature Conservancy Council (NCC) and the Large Blue Committee failed to conserve the insect.

It was the NC, which worked in association with the Large Blue Committee (an independent body with NCC represented on it), that oversaw the conservation plans laid out to save the butterfly. There was no question of the authorities not realizing that it was likely to become extinct and being taken by surprise, so it must be said that there was a positive strategy to conserve the insect. Indeed, it was the only insect protected by law since 1973, and there was a statutory obligation by the authorities to conserve it.

The Nature Conservancy set up a research programme in 1972 to study the Large Blue and much of this work was handled by the Institute of Terrestrial Ecology (ITE) which up to the establishment of the Nature Conservancy Council (NCC) was still part of the NC. Although the complicated life cycle of the Large Blue had been discovered early in the twentieth century by F. W. Frohawk, and the insect had been bred successfully in captivity by Cyril Clarke (later Sir Cyril) at Liverpool (Clarke, 1954; Howarth, 1973), a detailed ecological analysis of the species was assigned in 1972 to Jeremy Thomas of the ITE, the body charged with carrying out the NCC contract. **(2)**

THE INFLUENCE OF THE COMMITTEE ON LARGE BLUE CONSERVATION

The SCPBI tried very hard for a long time to buy a piece of land rich in Large Blues, and, after much trouble, finally leased land in Cornwall to establish a nature reserve for the Large Blue butterfly. This operation ran from about 1930 to about 1953.

That the Royal Entomological Society of London (RESL), through the SCPBI, tried to buy a nature reserve for a butterfly (the Large Blue) and eventually leased one for a number of years is an historic event for a society established 'for the improvement and diffusion of entomological science exclusively'. The RESL paid for 'watchers' during the collecting season and generally were more involved with butterfly conservation in these early days than they are today.

The search for a suitable place for the Large Blue started in 1926 and one locality at Millook was almost purchased, but fell through. Then the SCPBI tried a place near Hartland Point - this was Norman Riley's suggestion. Eventually one was found at The Dizzard, which is now owned by the National Trust. There were in fact numerous good localities for the Large Blue, including one called 'Butterfly Valley', this was 'Chipman Point (The so-called 'Butterfly Valley') a fact disclosed in R. A. Jackson's report to the SCPBI on the Large Blue in Cornwall, dated 12 July 1948. It is recorded in the Minutes of 1936 that the Committee still wanted to find land to buy in order to give it to the SPNR, but that the owner of a particularly good Large Blue site was not interested in selling.

On reading the complete minutes of the Committee one has the impression of difficulties of communication over a long distance; it was as if the Large Blue localities were somewhere in South-East Asia and that considerable effort was needed for anyone actually to visit and report back to the Committee. Finding out what was happening in Cornwall always seemed to be long-winded. Trips made by Committee members and others were very few. During and after the war years petrol was rationed which caused more difficulties.

Throughout the leasing of the Cornish coastal habitat, there was continuous discussion about the state of the habitat with respect to how it should be managed (by sheep, cattle, burning, cutting, or by no action), enormous amounts of time were spent debating the usefulness of noticeboards, the repair and maintenance of noticeboards, and discussion as to whether noticeboards were really a good thing drawing people's attention to the locality. There were at least two generations of noticeboards that weathered the westerlies of that region, whilst the fortunes of the butterfly hung on rather precariously along this beautiful coastline. One set of noticeboards deliberately excluded any mention of butterflies or the Large Blue, since it was agreed that it would otherwise draw unnecessary attention to the Large Blue if mentioned.

Conservation of the Large Blue

The Committee employed a 'watcher' most years during the Large Blue season to warn off collectors. Either the Committee paid the watcher's fees, or the RESL paid, 'providing the business was transacted in the name of Mr. Hedges, not the Society.' The watcher was employed from 20th June until the end of July at 36 shillings a week, for attendance 08.00 to 18.00 each day including Sundays. The actual cost of policing the Large Blue colony in 1931 was £13. 19s and 4d. Mr. Hedges was employed to do this job for 4 years. In 1934 the Committee took out a further 7 year lease of land at the Dizzard at the annual rent of £12 - to include the 'watcher', repair and winter storage of the noticeboards. In 1937 Mr. Hedges was paid off with £3 and he recommended leasing the site to a Mr. Sanguin.

At the end of ?1931, a Member (only mentioned as such) of the RESL sent a report to the Committee saying that the management suggestions of the Committee were working and the butterfly thriving. He recommended not cutting any vegetation at The Dizzard since, in his opinion, the thyme which grew below the gorse provided better shelter for the larvae once they had crawled down from the parts of the plant above the gorse. Time was ripe for another visit.

Miriam Rothschild volunteered to visit The Dizzard on the North Cornwall coast, since she was at that time working in the Plymouth Marine laboratories, where Spooner - a noted Large Blue enthusiast - was also working. It is stated in the 1946 Minutes of the SCPBL that fuel was eventually granted by the Ministry of Fuel especially for a trip to study the welfare of the Large Blues in the Dizzard area. Unfortunately Miriam Rothschild cannot now recall visiting the site nearly 50 years ago (pers. comm. 1994) but a report to the Committee by Spooner, who was accompanied by Freddie Russel, describes a visit these two made on 23 June and 14 July 1946. This four-page report in long-hand has a considerable influence on the conservation strategy of the Large Blue thereafter. Spooner's final paragraph is telling: '(viii) For the time being at least the whole matter of conserving arion in Cornwall is not one of urgency, as the insect has ample natural sanctuary (even if this is naturally growing .....?). There is time in hand to acquire more information about it. It is strongly recommended that one or more Cornish inhabitants should be persuaded to take special interest in the insect, preferably someone living near Bude.'

There were conflicting results. The Committee could not decide what sort of habitat was right for the Large Blue. It seemed to be existing at various places along the coast. Canon T. G. Edwards had just visited various places along the North Cornwall coast during July (?1946) and reported *arion* in 'six separate unfrequented valleys from Crackington to Clovelly a distance of 35 miles' This has been confirmed verbally by Graham Howarth (pers. comm. 1994) Labouchere had never seen the Large Blue so abundant. He checked Millook Valley - a favourite locality - but none was found there. Labouchere mentioned to the Committee that The Dizzard was now extensively overgrown. However there were these 35 miles (56 km) of coast where the Large Blue prospered. Weighing up all this conflicting evidence, the Committee were conclusive: 'It was finally agreed that no useful steps could be taken at the moment to protect arion in Cornwall, but that there was no cause for alarm as the insect had ample natural sanctuary for a considerable distance along the north Cornish coast.' From there on in the Minutes of the Committee, the welfare of the Large Blue slips to second place in favour of the Glanville Fritillary on the Isle of Wight. The Committee had been told that the Large Blue was seemingly doing quite well and that there was abundant coastal habitat for it - after all 56 km of coast seemed like more than enough. That some areas had become overgrown was not a real loss; there were plenty of other places.

The 'preserve' soldiered on until at least 1953 when Graham Howarth reported to the Committee that it was doing well and the noticeboards were still alright, the second set having been erected in June 1949. The Committee also tried another method in 1948 to conserve the butterfly, but in this they were unsuccessful; they tried, rather too hopefully, to sway the Cornish farmers into managing their land to suit butterflies. The Committee sent draft circulars to Mr. D. C. Thomas **(3)** of the National Agricultural Advisory Service in Exeter hoping they might inform the farmers, but he, eventually, convinced them they certainly would not.

Conservation of the Large Blue

HABITAT CHANGE

The gradual choking out of the Large Blue habitat by scrub must be put into perspective, since it was, and still is, the main ecological factor that everyone stated/s for the demise of the insect. In the 1950s the butterfly occurred virtually all along the North Cornwall and North Devon coast from Crackington to Clovelly. Some sites were already being overwhelmed by scrub (e.g. The Dizzard, although it was still OK in 1953) but there was room for complacency since there were plenty of other places that the butterfly was being reported in. By the late 1950s the virtual continual occurrence of the insect along the cliffs had wittled down to about half a dozen sites, and finally in a few years, to just one or two. This was co-incident with the arrival of the myxomatosis epidemic amongst rabbits, so the problem was then rapidly exacerbated.

The coastline at this time onwards then suffered from three environmental impacts, (i) loss of grazing through the loss of rabbits which led to the invasion of scrub, (ii) the intensification of farming which included the 'improvement' of marginal land and bringing rough coastal land into production (by ploughing), which had the effect of bringing the fields closer to the cliff-tops and restricting the areas where butterflies traditionally existed, and (iii) development of coastal areas as recreational, residential and touristic sites. These factors were typical of a nationwide trend, not just typical of the West Country, and impacted negatively on many coastal butterflies, not least of which was the Large Blue.

The rugged topography of the North Cornwall and Devon coast with its coombes and coastal cliffs must be emphasized as an important factor in the retention of the Large Blue in this its last outpost in England. The grassy slopes, valleys and grazed hillsides provided more than ample habitat for the butterfly away from the pressures of intensive agriculture, at least for as long as possible. The fact that modern agricultural methods and intensive farming came late to this region compared with the rest of southern England is one contributory reason why the Large Blue hung on there. Spooner, in his 1946 report, mentioned how out of the way the area was, and how he thought it would not be threatened because of it: 'The question of the development of this coast, approached only by a few abominable narrow hilly roads, does not at present arise and probably will not arise for a long time. It is therefore doubtful whether it is worth attempting protection for arion in this district: it has excellent natural sanctuary'. This was written well before the area became of tourist interest with long distance coastal paths. (4)

Conservation of the Large Blue

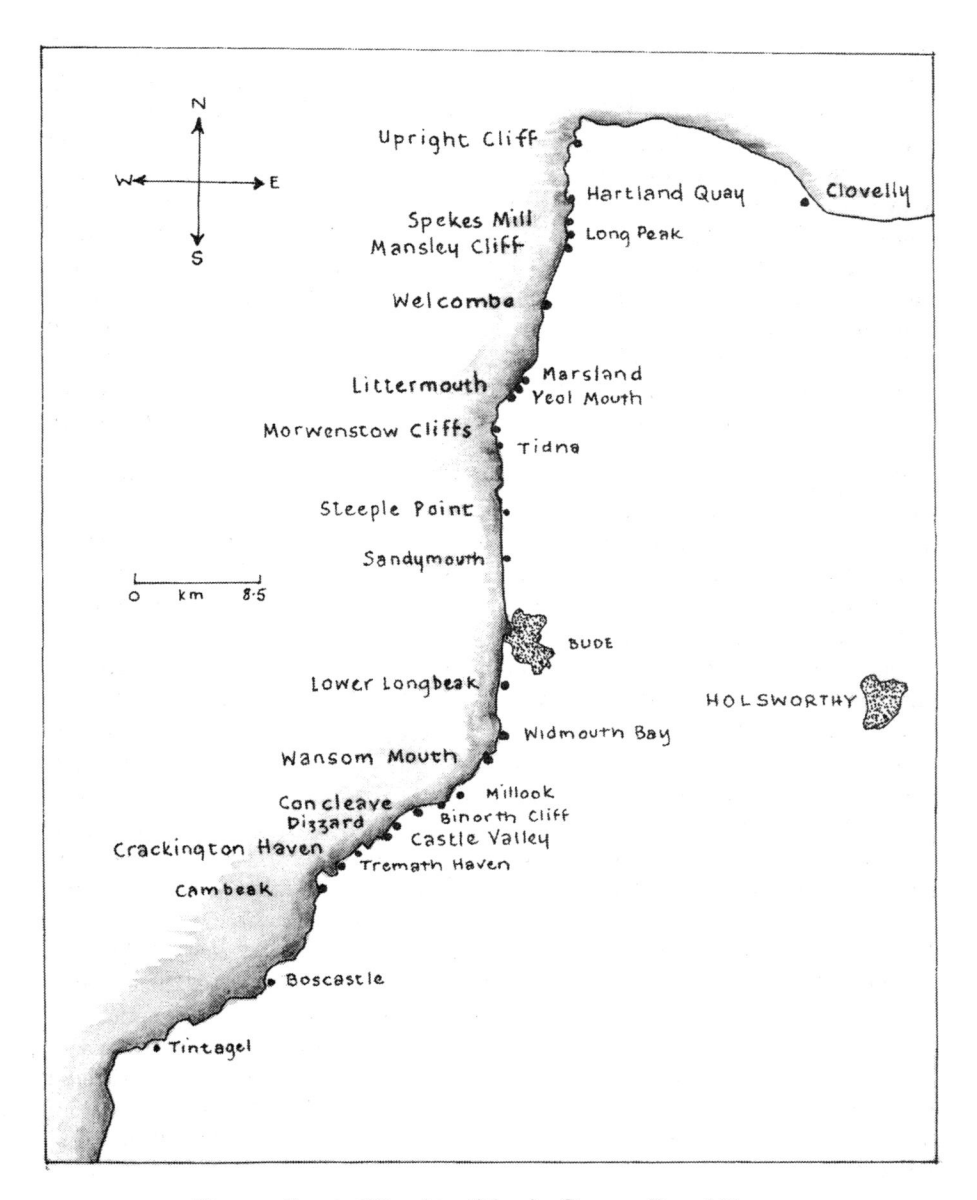

Former Large Blue localities in Cornwall and Devon
extended caption on page 96

Conservation of the Large Blue

THE ELUSIVE MR X

There may have been deliberate attempts to eradicate the Large Blue in Cornwall. Some might find the following evidence odd and unconvincing, but I will let the documents speak for themselves. Some of the Standing Committee for the Protection of British Insects (SCPBI) stated they had seen the evidence of deliberate extinction relating to damaged ant hills. Collectors were still incredibly active during this period. The information from the Minutes of the SCPBI states that a certain Mr X of the Royal Entomological Society of London (RESL) should inform the person carrying out destruction to the ant hills, and to draw attention to this person the gravity of the destruction. Why Mr X of the RESL had to remain anonymous is a mystery, unless he was involved too. In 1930 Mr. Sheldon raised the question of destruction of *arion* habitat in Cornwall: 'he (Sheldon) understood that efforts were still being made to exterminate the species in Cornwall, and the possibility of a protest to the individual in question was considered by the Committee' (Minutes of the SCPBI of 23 April 1930).

During 1931 there were five noticeboards placed around the Large Blue site, and the 'watcher' turned away 12-15 collectors. Again the wilful damage to the habitat of the Large Blue was raised at the Committee, but as there was no evidence and no names, no action was taken. Lord Rothschild had insisted that the 'watcher' should not only stop collectors catching adults but also prevent them from digging pupae from ant hills. The 'watcher' was Mr. Hedges.

One of the major workers in the field of butterfly conservation at this time was the Dutchman F. A. Labouchere who, having read a report about the ant hills being razed, visited the area and wrote a report for the Committee called *'Report on Visit to Bude'* dated 31 July 1935. It makes salutary reading: ..'I endeavoured to ascertain from the owner of one of the cottages at the foot of Millhook hill (*sic.*) (5) what collectors had been down there. She said that there had been a good many this year, and some of them had been very satisfied with their catches. The following are some of the names:- Messrs, Snell, Sutton, Sills, August, Colonel Wood, Bartlett, Rosborough and Greenwood. Of these Sutton and Greenwood had been in the neighbourhood for three or four weeks, and had collected assiduously. .. I understand from Mr. I. Heslop that some years ago he took 600 Palaemon.' Heslop was a commercial collector who worked with an assistant. As Labouchere reported: 'The place had evidently got known and had been thoroughly worked. It was trampled down in all directions, cigarette ends and match ends were scattered about, but the worst feature was that those ant-hills which were visible had been systematically opened up for pupae - the tops had been cut off and were lying alongside.'

As we know now the Large Blue does not breed inside ant hills, it breeds in the underground galleries of *Myrmica sabuleti*. The person who allegedly tried to exterminate the Large Blue from its localities might never be known, and the possibility that the deliberate vandalism might have been confused with active pupa-hunting cannot be ruled out. Once common in Victorian times, the practice of pupa-hunting was a widespread technique used to find pupae of lepidoptera. It may well have been pursued here, in ignorance, to find the pupae of the Large Blue; the usual place for pupa-digging is around the trunk of trees where the pupae of moths may be found.

OTHER ARION RESERVES

Enthusiasm for the Large Blue hardly waned. There were other attempts to re-establish the butterfly in Britain, some of these of a private nature. Perhaps the best example, and the least known, was done by none other than Mr A. H. Hedges. I am grateful to Graham Howarth for bringing to my attention the following information, which he recalls very clearly being told about by Dr. E. A. Cockayne. Mr Hedges is also the un-named 'rich person' in Howarth (1973) who let a Large Blue nature reserve become overgrown on the Devon coast. Mr. Hedges was by all accounts an exceedingly good breeder of lepidoptera, well known at exhibitions, etc., and sufficiently interested in the Large Blue that he bought a piece of land with a good colony of Large Blues on it. It was a cliff-top site complete with a coombe which he made into a conservation area. He operated this as his personal nature reserve and forbade anyone to go near it. Eventually the reserve became overgrown, Mr. Hedges left - he moved to the Isle of Man - and the reserve disappeared. Its location has not been determined. Mr. Alfred Vander Hedges (1893-1957) was elected a Fellow of the RESL in 1910 and remained in membership until his death. (6)

At another site, in 1958, there was an unsuccessful introduction of eggs of the Large Blue by a Mr. Murray, to his own private nature reserve at Otford (probably the one in NW Kent, nr Sevenoaks) which he had collected during the year in 'the Ordessa Valley, Spain, and in France' (Minutes of the CPBI, 15 January 1959).

Perhaps the most significant nature reserve purchased during this period for the Large Blue was in Devon during 1958. Through the generosity of Mr. J. C. Cadbury 65 acres (26.3 ha) of land on the south side of the Welcombe Valley near Hartland was bought and given to the SPNR for use as a nature reserve. The reserve was to be bought under the 'protective umbrella' of the Nature Conservancy (Moore, 1959). Most of this reserve lies within the Marsland Mouth SSSI.

An overview of these early attempts to conserve the Large Blue butterfly must be seen in the light of ignorance of its precise habitat requirements, some degree of complacency over the precise methods of management of the habitat, and a firm belief, engendered by Spooner, that all was doing well for this butterfly. Loss of localities because of growth of scrub was happening decades before myxomatosis, and was appreciated as a limiting factor. Norman Moore recalled 'two reserves were set up to protect the butterfly'... 'but neither succeeded in their purpose'. (Moore, 1987, p. 135). Incidentally, I have been unable to ascertain where this other reserve was. The Dizzard Valley Butterfly Reserve had been established in 1931. Thomas (1989) concluded that the Large Blue was 'killed by kindness' since, in ignorance of correct conservation management procedures, the site was allowed to become overgrown, owing to cessation of traditional farming methods over the area. Today the site is managed by the National Trust, grazed by cattle, and is famous for its dwarf stunted oak trees clinging to the steep-sided valley.

Norman Moore recounted the only time he ever saw the Large Blue 'feeding and flying on the south- facing side of a sheltered valley which ran down to the sea on the north Devon coast'. Sir Dudley Stamp (1969) recalled that he saw Large Blues on the north Cornish coast .. 'Butterfly Valley at Crackington Haven, Cornwall which had been carefully left untouched for thirty years to preserve the Large Blue, was so overgrown that it contained neither thyme, nor ants, nor butterflies'. That is a clear indication of non-management of the butterfly habitat. The butterfly's requirement for short turf has been recognised since at least the 1930s.

CONSERVATION, 1960 - 1979

Large Blue conservation picked up again in the early 1960s with the establishment of the Joint Committee for the Conservation of the Large Blue Butterfly, the JCCLBB, in 1962. This was originally established by the Devonshire and Cornish Naturalists' Trust but was later joined by most other national conservation groups. Despite their distinguished efforts over a decade the Large Blue declined to two colonies in 1972 and nil in 1979.

It was during this period that the NC also devoted a considerable amount of effort and resources into conserving the Large Blue; and they appointed Jeremy Thomas specifically to that task in 1972. During the period 1976-77 the Institute of Terrestrial Ecology (ITE) at Furzebrook (Dorset) were looking after one Large Blue site in the West Country; the total British population at this late stage was only 40 adults. Dr. Thomas advised the owner of the site about appropriate management of both it, and adjacent land, with the hope that the butterfly would extend its range. He was also hoping that sites where the butterfly used to occur could be 'rehabilitated', and that the species could be re-established there. That hope was dashed when the butterflies became extinct 12 months later.

These final years up to extinction are well documented by the NCC (1981). Most of the Large Blue populations in the Cotswolds, Somerset, Devon, Cornwall and Pembrokeshire had disappeared by 1965 through intensification of agriculture and the decline of rabbits. This was a dramatic change from Spooner's assessment and prediction in 1946, and Howarth's report to the SCPBI in 1953. The droughts of 1975 and 1976 reduced the population in the final colony in Devon, and bad weather reduced the feeble populations of adults in 1977 and 1978. 'In 1978 the five butterflies that emerged over an extended period were kept in semi-captivity and a batch of viable eggs was obtained. 'During the flight period in 1979 the same procedure was followed and the 22 Large Blues that emerged failed to produce any viable eggs. That was the end of the English Large Blues (NCC, 1981). In Table 7.1 summarizes the stages in its extinction.

Conservation of the Large Blue

Table 7.1 Countdown to extinction of the Large Blue

BEFORE EXTINCTION

1915 Full description of life cycle by Frohawk (1924)
1931 Dizzard Valley Butterfly Reserve (Cornwall) set up for the Large Blue.
1950s (early) 30 colonies left. (Habitat, 1979)
1962 Formation of the Joint Committee for the Conservation of the Large Blue Butterfly (JCCLBB)
1963 80 adults seen
1965 most colonies had disappeared
1966 Local farmer agreed to receive £10 an acre for not ploughing land on Large Blue locality in West Country. This is presumed to be first example of 'paying farmers not to farm' later adopted on a wider basis as 'set-aside' in the UK. (Minutes of Entomological Liaison Committee 22 March 1966).
1971 12 Large Blue sites left (Dempster, Note of 21 June 1971 to Ratcliffe, NC)
1972 Two Large Blue sites remaining.
1972 NCC research programme given to ITE
1975 Put on Schedule I of the Conservation of Wild Creatures and Wild Plants Act 1975 (Habitat, 1979)
1976 40 adults left in Britain (Morris, 1977)
1979 22 butterflies emerged, but no eggs laid
       NCC and ITE announce extinction of Large Blue (NCC 6th Report, 1980)

AFTER EXTINCTION

1981 Protected under Wildlife and Countryside Act 1981

REASONS FOR EXTINCTION IN 1979

Thomas made two discoveries just prior to the Large Blue becoming extinct (Thomas, 1989). First, agricultural practices had changed; there were fewer sheep and cattle grazing over rough ground especially on shales, which also had formerly been burnt off, or 'swaled'. This had previously produced the ideal habitat for the Large Blue. Second, the host ant, *Myrmica sabuleti* was dependent only upon these warm sheltered south-facing slopes in southern England. Thomas (1989) says 'These discoveries were made shortly before the final British Large Blue colony became extinct'. The involvement of the butterfly with the any species had however been reported earlier by Spooner, Moore and Howarth. Spooner (1946, p.3) had reported in detail on the ants of a suitable Large Blue locality in Cornwall, which included '*M. scabrinoides var. sabuleti*' as being part of the ecological association of the Large Blue habitat. Moore (1959) had also discussed two ant species on which the Large Blue were dependent including '*M.scabrinodis*'. Howarth (1973) also noted *M. sabuleti* as a part of the ant fauna typical of Large Blue sites.

Moore (1959) is quite unequivocal that the demise of the Large Blue was mainly caused by over-collecting - and that 'unless it is adequately protected in its principal locality, it will shortly become extinct in Britain.

In effect there were several factors which contributed to the demise of the Large Blue from England. Collecting is not generally regarded as damaging except in small populations, and Large Blue populations were small prior to extinction. There is no doubt, as Howarth (1973) stated, that collectors have major impact on populations when their numbers are low.

Collecting remains a distinct factor in the demise of the Large Blue and would have impacted negatively on surviving populations. It is difficult to prove either way, but collecting as a factor is disputed by Thomas who says that all the objective investigations into the effects of collecting have not shown it to be the case. He does admit that the Large Blues in Northamptonshire were, however, annihilated by collectors. Most would admit that any collecting on a small population is likely to have significant damage. Many sites were taken into intensive agricultural use by ploughing. Many sites were lost to afforestation: and Heath *et al.* 1984 recorded that half of the 90 recorded sites in Britain disappeared because of this. Myxomatosis played a significant part. The loss of rabbits meant that the short turf needed for the butterfly became overgrown. This was the final nail in the coffin of the Large Blue (NCC 1981). It was the release of the myxomatosis virus into the British countryside during 1953-55 that was the crucial factor, which, Moore admits, caught the NC unawares, taught them many ecological lessons fast, and was ultimately responsible for the demise of the Large Blue by indirect means. It is only in more recent times that the release of

Conservation of the Large Blue

organisms (modified or not) such as the myxomatosis virus into the countryside has been strictly controlled.

The final factor that possibly impacted negatively upon the Large Blue was genetic effects caused by inbreeding through isolation. The fact that the last living 22 adults of the English Large Blue did not lay fertile eggs suggests that some adverse genetic effects may have contributed (i.e. mating between genetically related individuals). Inbreeding can reduce useful genetic variation in a population and can result in the expression of harmful genetic traits. John Muggleton assessed all the available evidence drawn from populations of Large Blue from Northamptonshire to Cornwall, and especially in the Cotswolds (Muggleton, 1973, Muggleton & Benham, 1975) and argued the case for the possibility of genetic isolation having brought about the extinction of the Large Blue.

Thomas (1989) dismissed Muggleton & Benham's work saying that 'None of these theories stand up to any scientific examination of the facts', but no evidence was provided to explain this conclusion. Berry (1977) also postulated that genetic isolation was unlikely to be disadvantageous. (7)

That neither Muggleton nor Thomas either presented empirical evidence suggests that we will never know if there was any adverse genetic effect acting on the Large Blue. The only way that adverse genetic effects could possibly have been proved categorically would have been, if, at the time, Large Blue butterflies were analysed by molecular techniques to determine whether they were becoming more homozygous or not. They were not.

The preferred state of affairs for any butterfly, or indeed any invertebrate, is to have a large amount of variation (= high heterozygosity). They have a great reproductive potential and they can 'afford' to lose a significant proportion of their young, typical of many butterflies. The information gleaned from inbreeding work on other insects, such as the Fruit fly, *Drosophila*, shows that they are affected by isolation, whilst others are not, but then *Drosophila* populations are routinely derived from a single female. The same kind of inconsistent work has been shown with mammals and birds making any comparison with insects meaningless.

According to Hedrick & Miller (1992), research on fruit flies indicates that unfavourable genes are the major cause of inbreeding depression. A very small population of butterflies of any species forced to breed amongst themselves might fairly quickly throw up disadvantageous characteristics. This would presumably apply to any species low on numbers. The sorts of characteristics which might manifest themselves where adverse genetic effects occur, could be in their reproductive potential, courtship, finding mates or just about any physiological aspect of a butterfly's life which would be almost impossible to detect or monitor. Perhaps part of the problem lies in the population being stressed; but stress is very difficult to measure. As Berry (1992) stated (in quoting Calow, 1989) the best definition of stress 'is any environmental influence that impairs the structure and functioning of organisms such that their neo-Darwinian fitness is reduced'.

What is needed now is some experimentation on stress in butterfly populations. No one as far as I am aware has ever done this. It is possible that the Large Blue's decline was due to poor survival through stress produced by a deteriorating environment. The fact that precious few individuals existed in a dwindling population might lead one to suppose that they were at least under stress and that there was no 'new blood' on offer. Severe restrictions would therefore act upon its gene pool. The chances of inducing genetic breakdown runs higher, the lower the number of individuals in any population, theoretically, but this is all guesswork. With just 22 Large Blue butterflies left in 1978, it is likely that there were severe selective pressures operating. Maybe they already started in the previous season and accounted for the demise of the last twenty-two butterflies, all genetically unfit, and highly stressed. More work has still to be done.

The fact that the great mass of ecological data on the decline of the Large Blue from its continuous distributional range (virtually all the way along the scarp slope of the Cotswolds, north-east even to the Northamptonshire populations and south-west to Cornwall) was overlooked by commentators anxious to show that there were no adverse genetic effects, is to ignore, in some people's opinion, a lot of circumstantial evidence and the fine ecological history of this butterfly.

The public are perhaps not particularly interested in whether there has been genetic decline or not. They just want butterflies in their respective habitats. Perhaps the scientists will sort the precise ecological constraints eventually. The known facts is that the butterfly went on a steep decline to extinction. That the scientists did not save it after decades of work, appears to be a missed opportunity in the world of butterfly conservation.

The history of the demise of the Large Blue is one of increased isolation of populations, of a retreat from the east to the west, and a continued nibbling away at the western coastal strongholds until the species finally expired with no where else to go. As Muggleton & Benham (1975) pointed out, E. B. Ford (1965) calculated that the survival rate of adult butterflies (Meadow Browns) was lowest in the colony with the smallest area. Evidence is given by Muggleton & Benham (1975) about the apparent interdependence of adjacent colonies, and how adults can fly several kilometres. When there were pockets of suitable habitats close by, the insect could colonize these, in the same way that the Heath Fritillary colonizes new woodland patches. Thus the Large Blue appeared to live in patches, but at the same time was not entirely isolated in localities and was able to exist almost as a continuum along suitable habitats, such as the limestone grasslands or coastal cliffs; And it did so as a cline. The brighter blue butterflies in the most western populations which some people were eager to draw attention to.

Conservation of the Large Blue

We must conclude that we have no direct evidence either way about any possible adverse genetic effects although there is more circumstantial and theoretical evidence for than against. The demise of the Large Blue could simply have been caused by the butterflies flying away, prospecting other parts, and, finding them unsuitable, carrying on, never to return to their original localities. It is a more comfortable thought (to some) to believe that the Large Blue exists as a relatively neat 'metapopulation' and that any return to the original location is disadvantageous. Why would butterflies want to return anyway; it is in their nature to go abroad?

The explanation for the Large Blue extinction could be conceptionally simple. It was simply that the Large Blue did not have the range of genes in the first place to be able to make adaptations to a changing environment. The rate of environmental change through the intervention of humans, galloped ahead of the normal changes in the natural environment, for which the Large Blue has some genetic capabilities. Its range of genes had been limited perhaps by adaptation to a specialized environment. One must take a broad overview of the demise of the Large Blue. The lycaenids are one of the most successful butterfly families worldwide. They have become highly specialized, and therefore are at a disadvantage when it comes to habitat changes. The extinction of the Large Blue has to be regarded as a minuscule loss in the world of the lycaenids, and should be accepted as such.

That the demise of the Large Blue was 'inevitable' was dismissed by Vane-Wright (1977) as a 'stab at the heart of Darwinism'. This dismissal must be viewed with great reservation. For an insect that has such an intricate ecological niche, there is little room to manoeuvre when the conditions change. Change they did, courtesy of human intervention. Unfortunately the ecological pressures were heaped on these butterflies when their populations were low, when all sorts of artificial interferences were taking place, and finally their populations succumbed. There was an air of the inevitable. Many could see it happening.

If a graph depicting the tumbling decline in the populations of the Large Blue from early in the twentieth century to the 1970s was made one could have predicted that most Large Blue colonies would become extinct. Perhaps we are all the wiser in retrospect. Despite this Vane-Wright (1977) reported Thomas as saying that the Large Blue was not doomed to extinction. Twenty four-months later the NCC declared the butterfly effectively extinct in England.

The colonies in the West Country had disappeared, as Moore puts it simplistically, 'largely due to myxomatosis'. It is the delicate ecological balance between the Large Blue, its foodplant, its host ants, its state of parasitism, the state of management of the habitat and possible climatic affects that is central to attempts at re-establishment. No-one seems to have had a handle on all the ecological inter-relationships; and no-one knows all the factors now (David Sheppard, pers.

comm. 1994). Yet the butterfly slipped away even when it was afforded all the conservation the authorities and 'experts' could muster; and with almost 20 years research post-extinction, the butterfly is still not firmly re-established. This is perhaps an indication that a change in the climate may be influential.

OVERVIEW ON EXTINCTION

Jeremy Thomas's 1989a article on the Large Blue makes good reading. It is a forthright review of his work on the species, with a good overview of the historical background to the species. If there is just one Large Blue article to read, this is it. However it is a salutary tale, not one of overriding success. After two decades of research Thomas is still hoping to have the public enjoy the fruits of his teams' labours. The sites are, however, still secret, and despite the best of financing from government, private companies and conservation charities - i.e. the very best that can be afforded to any single butterfly species, or any British species - it cannot be said to be an amazing success. In fact it has been, and continues to be, a failure (Thomas 1984, 1994) despite everyone's best hopes. The Large Blue story still goes on. But what of the results? One is left with the feeling of long-term, dissipated energy; has it all been worth it?

Could the butterfly have been saved? Possibly. Seventy years before Thomas accepted the job of conserving the Large Blue it had been bred on three separate occasions in captivity by Purefoy (Howarth, 1973), and then more recently by Cyril Clarke (Clarke, 1954). When the butterfly was thought to be on the edge of extinction in 1970 the Large Blue Committee asked Robert and Rosemary Goodden to investigate the feasibility of breeding the species in captivity; despite experiments with French stock collected in the Alps, this avenue proved to be unsuccessful at first (Goodden & Goodden, 1973) but was later successful, although hugely complicated (Goodden & Goodden, 1974). Commercial production on a large scale however, appeared out of the question. Stock was not apparently taken from the remaining English colonies by anyone, even the ITE, in the years leading up to extinction, to ensure its safety.

So, have the British public got value for money in this government sponsored exercise to conserve the Large Blue? With the conservation of other important butterfly species in England there are some results to show for all the hard work and resources spent. Wicken Fen is of great international importance for many things, but in the British context it is very important for the conservation of the Swallowtail, and the important wetlands at Woodwalton Fen support the Large Copper. For the Large Blue there is nothing tangible, only continued experiments at Site X and Site Y, and nothing for the public (except there have been some unofficial coach parties to the 'secret reserve'), although the public have

Conservation of the Large Blue

contributed more to this butterfly than any other. Critics might ask, if the autecological aspects were learnt just too late to save the butterfly, why has it taken another 20 years to have a butterfly that is still at such fragile numbers. It is clear that we still do not know all there is to know about the Large Blue. If we understood its complete autecology then it would be much easier to conserve it. When will autecological studies be completed? Jeremy Thomas is still under contract (via the ITE) with EN to re-establish the Large Blue; he is also much involved with parasitoids of the genus *Maculinea* in a European perspective which is likely to lessen the focus on the English Large Blue.

The justification for the continued energies on Large Blue conservation can only be that it is furthering our knowledge of how best to re-establish or introduce species. The fact that the insect is extinct is a powerful reason for trying to re-establish it again on the British list, and there is now a historical factor operating, that the Large Blue is a focus of endeavours to conserve butterflies in general. There must be enthusiasts of other species worthy of the limelight. The Large Blue remains a focus of attention with the Species Recovery Programme.

Much of Thomas' work has been on the autecological studies of the Large Blue, and was done before he embarked on a conservation strategy for the species. Unfortunately it was done too late and, as is always the case, did not reveal everything that there was to know about the insect and its associations. A rather thorough assessment of the botanical indicators of typical Large Blue sites was published by John Muggleton showing that there 11 wild plants were typical of all Large Blue sites (Muggleton, 1973). This work has never been properly tied into Thomas' work for its important ecological implications.

Dr. Muggleton's work, however, ties in very closely with the observations of E. B. Ford who drew attention to the fact that the Large Blue was known from 'the most varied of habitats' (p.126) such as the rough low-lying ground of Barnwell Wold (Northamptonshire). We must not get confused with the stereotype of the kind of habitat where it was last seen, such as the coastal slopes of Bude, or indeed the short-turved habitats that Thomas studied, as if this is the only type of habitat to be re-created.

Conservation of the Large Blue

Caption to figure: The eleven plant species shown by Muggleton (1973) to be associated with Large Blue localities: *Brachypodium pinnatum, Bromus erectus, Festuca ovina, Carex flacca, Lotus corniculatus, Cirsium acaule, Hieracium pilosella, Poterium sanguisorba (=Sanguisorba minor), Helianthemum chamaecistus (=H. nummularium), Plantago lanceolata, Thymus drucei (=T. praecox)*

Conservation of the Large Blue

Since autecological studies on the Large Blue are on-going, and since repeated re-establishment since 1972 still has not produced habitats overflowing with butterflies, we much conclude that there may be something seriously deficient in our knowledge of the way of life of this insect, or of our management of it.

Much of the intricate ecology of the Large Blue was already known from the works of T. A. Chapman, E. B. Purefoy and F. W. Frohawk between 1916 and 1953, and as Thomas says 'it is astonishing how accurate these were 'although they did not closely define the ant species, the larval behaviour and the habitat requirements which became fully known only after extinction. He confirmed nearly all these early workers' observations, 'save one small but important omission'; that the larva falls off and waits under a leaf to be adopted by an ant (in the wild), rather than walking away (in semi-captivity) (Thomas, 1989a). This is rather an important conservation point since management for the correct density of ants nests is vital to having enough ants to find the larvae, since the larvae do not find the ants.

Unfortunately Thomas had an unenviable task: sorting out the unexpected ecological imbalance exacerbated by myxomatosis which had been brought to England 20 years earlier. Scrubbing over of habitats was a major problem considered in detail by Lord Rothschild in the 1920s at the Committee for the Protection of British Lepidoptera, 50 years before Thomas started work on the Large Blue.

The verdict on the conservation of the Large Blue is that everyone who had played a part in its protection, including a dozen entomological committees (Chapter 2) who presented their best ideas, failed to conserve this British insect, and this included the important conservators (NC/NCC), the ITE, the Large Blue Committee and the SCPBI before it. The Large Blue was declared 'probably' extinct on the 12 September 1979 at 0001 hrs by the NCC in the form of a Press Release (NCC, 1979). We will have to wait until 1979 plus 50 years for it to be definitely extinct in Britain.

CONSERVATION POST-EXTINCTION

Conservation post-extinction is based almost entirely upon a re-establishment programme. Energies and resources have been spent on introducing Large Blues from just about anywhere in Europe that could supply stock from the wild. Many parts of Europe proved to be severely depleted in larvae, according to the journeys of Thomas (records at English Nature) and a colony in Sweden was eventually chosen as a source from which Large Blue stock could be introduced.

If one feels that *Maculinea arion arion* was an entirely British subspecies, then there is no question of re-establishing this, since there is no-where else in the world where this (now extinct) sub-species exists (even the more colourful cline that occurred in the West Country could not be materialized from anywhere). The Large Blue conservation programme is therefore a re-introduction programme, rather than re-establishment programme, since it involves introducing other Large Blue sub-species which are nothing to do with the original subspecies of the United Kingdom. Some Continental subspecies actually breed on plants other than thyme, and, more fundamental than that, each subspecies has evolved a unique harmony with its local environment and is in total ecological synchrony with those habitats. Some ecologists would argue that it would be a futile exercise to attempt to introduce species or subspecies into a new country since they would be ecologically unfit.

On the question of perhaps the most appropriate subspecies to re-establish or introduce into England from the Continent, there are only a few subspecies which would appear to be appropriate. Although there are 11 subspecies of *Maculinea arion* in France (**8**) there is only one that is very similar to the English subspecies, and that is *Maculinea arion eutyphron* Fruhstorfer which follows an original description by Oberthür of a specimen from 'Angleterre, Cornwall'. This subspecies is the one closest to the previous West Country populations of the Large Blue, namely on the north-west coast of France in the region of St Malo (Ille-et-Vilaine) - separated now by about 258 km of open water. As reported by Roger Verity in 1952 *'il y a une trés grande conformité de l'arion en Grande-Bretagne et en Ille-et-Vilaine'*. There has never been an official attempt to introduce this subspecies from this part of Brittany, or any other subspecies from anywhere else in France, into England as far as the authorities in both France and England are aware. Such an obvious first starting point might have been overlooked.

A worrying comment is put at the end of the NCC's report, that they would consider re-introduction from the 'Continent of Europe' 'when it is clear beyond doubt that there are no other colonies in Britain' (NCC, 1981). If both the British and Continental stock were identical genetically then there would, presumably, be no need to wait. The NCC therefore presumed, on advice from the Large Blue experts perhaps, that the Continental subspecies was genetically different! If they were the same then there would be no hesitation in introducing the butterfly from the Continent.

Despite this, a re-establishment scheme went ahead, and to the present day is still steaming ahead on a policy of re-establishing the butterfly into its former haunts in England.

Piecing together information on re-introduction progress has been hampered by the facts never being disclosed properly. At the end of 1979 Derek Ratcliffe was hoping that someone might artificially re-introduce the Large Blue and keep it going (Ratcliffe, 1979). The intensive ecological studies had been done (see

Conservation of the Large Blue

Thomas's numerous publications) and the work of Frohawk had been supplemented. Now it was time for conservation. But it was all too late.

In 1984 with the publication of the *Atlas of Butterflies in Britain and Ireland* (Heath *et al.* 1984) it was still only a possibility that continental stock could be introduced. 'Good conditions have been recreated on a few sites', so here we perceive that conservation was at last starting, albeit belatedly.

In 1986 the *New Scientist* reported that Thomas had successfully completed a 3-year trial of raising Swedish Large Blues at a secret location in the England, and that it was time for a full-scale introduction (Jones, 1986). The Large Blue became a national celebrity hitting the headlines in all sorts of papers and magazines. The actress Jane Asher promoted the Large Blue introduction from Sweden (the island of Oland), which was financially supported by Holland & Barrett and the World Wildlife Fund UK. Hundreds of store lines in the Holland & Barrett shops carried information on the venture, and a contribution was made to the £20,000 appeal. Butterfly stamps and cards were given away.

A full account of the various attempts at re-establishment (at the un-named secret location) are outlined in Thomas (1989) up to 1988 and are summarized in Table 7.2. The various items mentioned in the table are gleaned from publically available documents, but there is a dearth of reliable first hand information on which to judge the re-introduction programme because of the continued secrecy of the operation.

Since the early 1990s the Species Recovery Programme has taken over all the conservation plans of the Large Blue including introduction from the Continent. Fourteen organisms in Britain (already protected by the Wildlife and Countryside Act 1981, Schedules 5 and 8) have been especially selected to have this privileged treatment. An extension of the scheme now allows the English Nature to consider Red Data Book (RDB) 1 and RDB 2 category species for a 50% grant. Therefore about 1000 species are eligible. Few British butterflies are rare enough for consideration.

It does seem rather a misnomer for the Large Blue to be on the Species Recovery Programme, since *Maculinea arion arion* became extinct in 1979. English Nature has committed itself to 'recovering' a British subspecies with a non-British subspecies, but so be it. English Nature support the work and The National Environment Research Council (NERC) give grants for such work, but they do not control individual projects within ITE The latest document on the conservation of the diversity of the flora and fauna in Britain still concentrates on the Large Blue butterfly as a target organism giving precise aims and objectives (RSPB, 1994). At least the survival of this extinct butterfly - or another country's replacement is assured (Appendix 1).

Table 7.2 Re-introduction data on the Large Blue

Date	Comments
1976	Conservation management of Site 'X' and Site 'Y' continues; Site 'M' at St Catherine's Tor abandoned (JCCLBB, Draft paper 24 July 1976).
1977	Site 'Y' likely to be ready to receive butterflies (JCCLBB, Draft paper 24 July 1976)
1979	Site 'Z' should be ready within 4 years to receive butterflies (Large Blue Management Committee Meeting, 20 February 1979).
1982	2,500 Thyme plants established in Sites 'X' and 'Y' in a National Park in Devon.
1983	Feasibility project: 97 larvae taken from Olau Sweden and released at English site; emergence on sites monitored, survival rate investigated. Adults from Sweden introduced into UK (Thomas, 1986 see below)
1984	Seven adults emerged from previous year's introduction; site visits made to France, Belgium, Sweden and Poland; Old Winchester Hill and St Catherine's Tor investigated as possible re-introduction sites.
1985	'Small emergence'
1986	'Emergence doubled to ... dozen adults' (JCCBI, 1987) more Swedish larvae released at site * Several hundred eggs from Sweden flown back to ITE Furzebrook, and these produced 200 larvae which were scattered over a Devon site (Brassley, 1990)
1987	'Roughly 75 adults emerged' at the site in Devon and these laid ca. 2030 eggs (JCCBI, 1987, Brassley, 1990)
1988	150-200 adults emerged in Devon site (Brassley, 1990) 'nearly 200 adult butterflies' (*sic*) laid ca. 4500 eggs Second locality started in West Country, 'by translocating 22 adults and a number of larvae' (Oates & Warren, 1990); these were from the Devon site (Brassley, 1990).
1988	16 females and 43 larvae introduced into Somerset site; site managed by Somerset Trust for Nature Conservation (Thomas,1990)
1989	Guesstimate of 60-100 adults (Oates & Warren, 1990), laid ca. 3000 eggs Somerset site produced about four sightings from 1988 introductions (Thomas, 1990)

Table 7.2 cont..

Conservation of the Large Blue

1990 onwards.... By this date there had already been introductions of Swedish Large Blues into two sites in Devon (Sites X and Y), one into a site in Somerset, and one into a site in Gloucestershire. One of these will eventually have public access, one site is already a County Trust nature reserve, none is a NNR. As long as management agreements are present, there is no greater security than if the sites were NNRs. The 5-year aim of the Species Recovery Programme is to re-establish the butterfly on six sites. There are a further three sites in Gloucestershire, one in Somerset and one in Cornwall where management is in progress (1994) to bring these sites into condition for re-establishment (Sheppard pers. comm. 1994). In a letter dated 13 March 1986 to Ian McLean at NCC, Jeremy Thomas stated that about five adults had been introduced from Sweden over 3 years.

The 50% decline in the established population in the 1989 was due to the severe drought that occurred that summer. No further details are available after 1989.

* A document dated 23rd May 1986 from the Swedish National Board of Agriculture (Plant Health Division) to the NCC (and inspected by me at English Nature Headquarters), confirms that the Swedish authorities required no export permits for *Maculinea arion* although 'the species is endangered in our country'. It is a fact that not all Large Blue species have been designated, even though a motion to have all *Maculinea* species included in December 1992 did not win universal support. (Lhonoré, pers. comm. 1995).

Notes

1. The Large Blue was officially declared extinct on 12 September 1979 by the NCC and ITE (NCC, 1981). The extinction of the butterfly (rather than the announcement thereof) is usually referred to as 1979 by most authorities, but the NCC date of 1978 (NCC, 1981) is clearly an error. The significance of the extinction was that this was the first of the eight organisms protected under the Wildlife and Countryside Act 1975 to have become extinct. For a short while after 'extinction' there was active correspondence in The Times about a secret population of Large Blues in the Cotswolds, but this has since been discredited.

2. The ITE was established in 1973 to study the factors determining the distribution and abundance of individual plant and animal species, and the structure and functioning of terrestrial ecosystems and their interaction with freshwater. This research is aimed at providing a sounder scientific basis for forecasting and modelling the environmental impacts of natural or other changes, such as pollution or changes in land-use. The results of these studies are available to those responsible for the protection, management and wide use of our natural resources in Britain and overseas (NERC, 1987).

3. No relation of Jeremy Thomas (J.A.Thomas, pers. comm. 1994)

4. In the case of North Cornwall, the development of the 'Cornwall North Coast path' would later bring more people into the area than had previously been the case. The concept of 'Heritage Coasts', of which South Cornwall took a part in 1976, was an initiative of the Countryside Commission endorsed by the government in 1972.

5. Millook was a former rich habitat for the Large Blue, now shown on Ordnance Survey maps as 'Millook' and 'Millook Common'; it is otherwise spelt "Millhook" in earlier literature. Today it is one of the few valleys on this stretch of coastline that is covered with wood or scrub along its entire length of about three km. It would have provided superb habitat for the Large Blue in the past when it was more open.

6. H.B.D. Kettlewell records that Hedges was 'without doubt one of the foremost breeders of British Lepidoptera we have ever had, and his collection consists practically entirely of bred specimens (Kettlewell, H.B.D.K., 1958. Obituary, A.V.Hedges, The Entomologist. 91:.51).

7. I am very grateful to Professor Sam Berry of University College London for his comments on this section.

8. I am grateful to Professor Jacques Lhonoré of the University of Le Mans, France for drawing my attention to this information which may be found in Beaufort, F. de. & Maurin, H. (1989) Utilisation des Inventaires d'Invertebres pour L'identification et la Surveillance d'Espaces de grand Interet faunistique. Published by the Secretariat de la Faune et de la Flore, Paris and in Roger Verity's 1952

Conservation of the Large Blue

classic title: Les Variations geographique et saisonnières des Papillons diurnes en France. 363pp. Charles Oberthür (1845-1924) was a noted collector of butterflies and lived in Rennes in Britanny. Oberthür's important collection, including the English Large Blue, was purchased by the NHM which is where it now resides.

The Large Blue

R.I.P. 12th September 1979

Caption to Map on page 77
Map of North Cornwall and North-west Devon coasts indicating the historical Large Blue sites. The Large Blue used to occur extensively along the coast from between The Dizzard in Cornwall to Clovelly in Devon. Much effort was made between 1925-1953 to maintain the populations along this coast, but finally they were left very much to themselves. The RESL played a major part in Large Blue conservation, since it had a reserve here, through its *Standing Committee for the Protection of British Insects*. Many of these localities are mentioned in the text, or are of specimens labelled with these localities in the Natural History Museum. One of the purposes of this map is to give visitors to the former Large Blue 'ecosystem' a chance to see for themselves such a collection of classic sites, the unusual topography, and to soak up the ambience of the kind of habitat in which the Large Blue flew, to understand the succession of nature - from open grazed areas to thick stands of vegetation - and above all to marvel at the uniqueness of this butterfly which existed in this extreme oceanic climate from the foot of the Atlantic in the Mouths and Havens, up the steep-sided coombes.

CHAPTER EIGHT
Habitat Management

'Butterflies are one of the most threatened groups of animals in Britain' Warren, 1991a

'...although the management of the vegetation is vital to the survival of many butterflies, it is the subject about which researchers and managers know least' Warren, 1991b

'Habitat patches act as terrestrial archipelagos' Thomas, C.D. *et al.* 1992

It must be said that conservators are only just feeling in the dark, with regard to the habitat management of butterflies. If the autecology of the Large Blue is still not known after all too long trying to work it out, then objectively there is very little hope that we will be able to find a sinecure for the remaining butterflies in Britain. Habitat loss continues whilst butterflies are counted and

Habitat management

scrub is cut. Are we all fooling ourselves that in fact habitats can be managed properly for butterflies? Habitats may be managed successfully, but it is not easy and habitats are often more complex than anyone realizes.

What is abundantly clear is that a lot of butterflies are in fact specialists at exploiting a mosaic of habitats rather than exploiting a single identifiable habitat. Many have multiple ecological requirements such as a combination of woodland margins, glades and rides (almost making them generalists, or 'generalist specialists').

It is as if these butterflies were connoisseurs of an artificial environment; the problem being that we are changing the environment faster than butterflies can make adaptations. Many butterflies thrive on a regularly rejuvenated habitat which offers a moveable feast of intricate habitat requirements often difficult to define. Before humans, this was provided by relatively slow natural events, now 'the environment' is seriously exploited by humans and the butterfly species of today are those that have the best adaptations to meet those severe ecological changes. As Erhardt & Thomas (1991) put it '20% of all British butterfly species are restricted to very warm, successional states within their biotopes' - and these are often created by humans. Species that feed and breed in gardens, orchards and waste areas are typical examples, as well as those associated with forestry, such as the Heath Fritillary, or agriculture, such as the Small Tortoiseshell. Armed with this sort of information the conservation strategy for butterflies - how their habitats are managed - tends to have evolved a sort of policy that 'lumps' groups of butterflies into their very general habitats such as chalk grassland, or woodlands, without being specific. By using such a policy are we taking chances with butterflies, gambling with such generalisations, without knowing the facts first?

The fine ecological requirements of certain butterflies make habitat management acutely difficult to administer, plan or predict. Erhardt & Thomas (*loc. cit.*) concerned themselves with the 28% of the British grasslands butterflies whose populations have declined whilst their foodplants remain extant. These include species like the Adonis Blue and Chalkhill Blue which 'react very quickly to abandonment or intensification' of agriculture.

Perhaps the most worrying factor is that some butterflies require different things in different regions of Britain. This is seen with some of the larger fritillaries, and is manifestly obvious for those species that have different foodplants in slightly different habitats. In a study on the Scotch Argus, Paul Kirkland appraised the ecology and behaviour of its remaining English colonies and stressed that careful management was essential for its survival. He pointed to the fact that the colony in Grass Wood (Yorkshire) went from being abundant to extinct in less than 50 years (Kirkland, 1995). The Silver-studded Blue in North Wales is virtually restricted to south-facing slopes which are 7° C warmer in spring than north slopes, and to areas of short turf which are 8-13° C hotter than nearby tall vegetation. Such areas take about 5 years to develop after burning or cutting. This is very different from Devon where the same species breeds in heathland habitats which grow through the 'building' and 'pioneer' stages of development in 5-15 years (Erhardt & Thomas, 1991, citing various other authors). The constraints against finding suitable criteria (or constant conservation denominators) for each species thus compound. Because of this we need to conserve many individual ecosystems, which is even more difficult than conserving individual habitats.

The state of autecological research up to 1981 was that only eight butterfly species had ever been studied in this way. Thomas (1981b) reviewed the status of three, the Black Hairstreak, Swallowtail and Large Blue. The Large Copper had been extensively studied since the 1920s, Martin Warren had already commenced his work on the Wood White (Warren, 1981a) and Ernie Pollard had carried out work on the White Admiral (Pollard, 1979b). Much work had been done on the Marsh Fritillary, and Ken Willmott was taking a great interest in the Purple Emperor. There is, however, a lot of difference between autecological studies and conservation management.

Autecological research is all very well, it tells you (hopefully) what the ecological constraints are in the life of that butterfly but these studies are not always helpful for appropriate conservation methods, and some indeed, do not venture any comments on conservation. We are grievously lacking in vital ecological data, and are decades behind formulating any sensible conservation strategy, for most of the British butterflies. We might have to gamble before it is too late. However, the information we have learnt from the few can be used with reservation for the majority. We have to consider that butterflies that live in the same sort of habitat such as grasslands or woodlands could be conserved in the same manner - it is a big hope, but then these are the only data to go on.

The subject of habitat management is almost an impossibly large one to handle under one heading, but this chapter seeks to present an overview with comments on how management is progressing with various species.

A good place to start with discussion on habitat management is with some clear facts. Martin Warren reminded us that 'Nearly three-quarters of the 57 resident British butterflies regularly breed in woodland and all but five rely on open spaces such as rides, glades, or clearings' (Warren, 1991b). The highest priority for conservation he saw was for the 17 species that are more or less confined to woodland habitats over much of their range. Amongst these are the High Brown Fritillary and the Heath Fritillary which are the most endangered of all butterflies since they colonize the early successional stages of woodland. The woodland margin is very important in the mosaic of habitats that these woodland butterflies seek, as well as glades within woodland. Modern forestry

Habitat management

practice does not always make considerations for butterflies but there are many exceptions. In ride management of coniferous woodland there are good parameters already worked out, such that rides should be one and half times the height of ride-edge trees (Warren & Fuller, 1990).

Bernwood Forest is a good example where some of this habitat management can be seen in practice, and at least it tries to maintain the former richness of the site. (1) It is not only the width of the ride that is important, conferring both 'openness' and 'sunnyness'; the open 'box-junctions' at ride intersections are also good for butterflies. Box junctions are where the trees and scrub are cleared from an combined area of up to 50 m along each ride from the ride intersection. The result is a huge grassy glade sheltered by trees, and inter-linked to other parts of the woodland by wide rides. The vegetation that grows in the box-junctions is a mixture of grasses and other wild plants many of which are exploited by butterflies as foodplants. These open areas need to be managed to be kept open. Part of the success of Bernwood has been through the involvement of Butterfly Conservation who have assisted with conservation measures. As a demonstration of how to manage woodlands Bernwood is worth visiting as it has public access. Monks Wood (Cambridgeshire) is a similarly-managed woodland, but it is less easy to get access.

In the case of the woodland butterfly, the Black Hairstreak, Jeremy Thomas found that this was 'a relatively easy species to conserve' (Thomas, 1981b). What had been found in the autecological studies was that the butterfly lived in discrete areas within woodlands, and that so long as these areas were left alone when the wood as a whole was felled or otherwise managed, then the butterflies did very well. Attempts at making optimum habitats for the Black Hairstreak continue.

Work on the declining Silver-studded Blue (which inhabits limestone grassland, heathland and mosslands) indicated that it lives in very transient habitat: 'early successional stages at low altitude, predominantly on sheltered southerly facing slopes' (Thomas, J.A. 1985) especially in north-west Britain where Thomas carried out his studies. The butterfly lives in closed colonies and cannot fly far. The problem in its conservation is to keep its habitats open (ideally through grazing) and to maintain linkage between suitable sites to minimize distances between old and new habitats. Conservation solutions recommended included buying reserves for the butterflies, and translocating butterflies to suitable sites. There have been a number of successful releases in Wales, the best perhaps being at Rhyd-y-Foel in 1942 when 90 specimens from the Great Orme were liberated from which several colonies have been founded and populations of tens of thousands of individuals resulted.

Failure of the Lulworth Skipper to re-colonize isolated patches of suitable vegetation (grasslands dominated by its larval foodplant, Tor-plant, *Brachypodium pinnatum*) has been demonstrated by the re-evaluation of survey data by C. T. Thomas *et al.* 1992. Chris Thomas has been much concerned with the way patchiness of habitats is of use to butterflies. That butterflies use patches of suitable habitat ties in logically with the general way that they tend to use an ever-changing, fragmented and artificial environment. Thomas *et al.* 1992 also measured the relative patchyness of the Silver-spotted Skipper, the Heath Fritillary, the Black Hairstreak and the Silver-studded Blue.

Some very precise ecological data are now known about some of these species, for instance that the Heath Fritillary takes about 2-3 years to colonize isolated patches of suitable habitat which are only 300-700 m from the source populations. The exploratory flights of this butterfly, which is an expert at following the exploits of humans in an ever changing woodland situation, are modest, but all that is needed. Helping with the provision of corridors and rides linking suitable sites is always useful. As C. T. Thomas *et al* say, 'The patterns of occurrence are consistent with a stepping-stone model'. Black Hairstreaks were found by these researchers to be unable, in places, to exploit a shifting mosaic of habitat even though it was a few hundred metres from source populations, although they colonized new banks of Blackthorn, *Prunus spinosa* as they became available elsewhere. As for the relative colonization rates of translocated butterflies, it was found that the Silver-studded Blue managed to move 2.5 km in 41 years since release, and the Black Hairstreak 4 km in 36 years. C. T. Thomas *et al.* (1992) found that the more isolated the patch, the less likely it is to be colonized, and the greater the gap between habitat patches, the longer it takes for a vacant patch to be colonized. This might be expected for those species that have poor dispersal flights.

In his studies of the metapopulations of the Marsh Fritillary (which has suffered a 62% decline since records began and whose sites are declining at 11.5% each decade) Martin Warren (1994) showed that almost 50% of the colonies surveyed in 1983 occupied very small habitat patches of less than 2 ha in size. The Marsh Fritillary is a very sedentary and colonial butterfly which is not a great wanderer. Conservation of this wetland species might involve conserving a group of small patches of suitable habitat all linked to each other, so that a local population can benefit. The loss of Marsh Fritillary sites across most of eastern England may be explained by the break-up of metapopulations of the butterfly through the severe habitat destruction that the region has suffered. Often the problem in conserving wetland species like this is that drainage of the surrounding countryside lowers the water table making it difficult to conserve small habitat patches.

Learning fast from the conservation policy of the Large Blue - particularly the lesson that it is best to try to understand the minutia of the insect's ecology

Habitat management

before serious conservation commences, Martin Warren also set about discovering the habitat requirements of the Heath Fritillary. He was entirely successful in his studies, and was able to introduce a conservation strategy that was geared to expansion of populations. Managing the butterfly by managing the habitat was essential, and it was quickly appreciated that the butterfly exploited only young woodland glades following coppicing. Some colonies increased in numbers owing entirely to conservation measures implemented just in time. Whilst some woodland colonies of the Heath Fritillary still became extinct during Warren's studies, others were discovered in Exmoor. For further information on the management of coppice woodlands, readers are referred to Fuller & Warren, 1990. From a legal point of view most of the surviving colonies of the Heath Fritillary are now conserved within the framework of SSSIs

Where deciduous woodlands have been replanted with conifers the management for butterflies is seriously limited to the rides and woodland edges. What is good for some woodland species cannot always be applied generally in all woodlands. For instance Morgan reminded us that planting of conifers has been blamed for some of the destruction of Silver-washed Fritillary colonies, but in Carmarthenshire (Wales) this has actually favoured this species, in place of the Dark Green Fritillary (an open-ground butterfly) which has been deprived of suitable habitat (Morgan, 1991).

Topography and the amount of open or rocky ground are also very important in any woodland with regard to butterfly numbers in and around woodlands, especially for the larger fritillaries. Marney Hall *et al.* (1989) provided guidelines for modifying the management of coniferous plantations for butterflies and noted parameters such as width of ride, shade and abundance of important plant species.

The study of parameters for conserving butterflies in woodlands continued with the work of Ken Willmott on that gem of the English oakwoods, the Purple Emperor butterfly. Willmott's research stems from the 1960s and involves the conservation of other woodland butterflies including the White Admiral. Associated Tyre Specialists must be congratulated for sponsoring the work of Ken Willmott for the World Wide Fund for Nature, on the work he carried out on the ecology and conservation of the Purple Emperor (Willmott, 1987). Unlike a lot of reports on butterflies, the Willmott report combined both ecology and conservation in his report. It is rare for the conservation aspects to be discussed in such detail in a report, but then Willmott had, up to 1987, twenty-five years studying his insect in the wild. As it happens, Willmott lived at that time close to a fine Surrey locality for the insect, not far from one of Moses Harris's (1731-1788) Emperor localities at Kingston-upon-Thames, Surrey. Habitat improvement for the butterfly is at the forefront of the

Habitat management

Willmott report, and he would have us all planting Crack Willows (*Salix fragilis var. fragilis*) and Sallows (Goat Willow, *Salix caprea*, and *S. cinerea*) everywhere for the Purple Emperor.

Having a keen eye for detail (and for the early stages of the insect) Willmott identified the precise habitat requirements of the insect - key groups of trees, master oaks, gaps in the canopy and young growth of Sallows especially ones with young succulent leaves. These criteria were part of the habitat structure that Willmott proposed that those who manage woodland should try to ensure for Purple Emperors. Ken Willmott has published several reports on the conservation of various woodland butterflies for the National Trust (NT) and the Forestry Commission (FC) but these are in-house documents not available to the public. It is very reassuring that large parts of the North Downs, and parts of the Surrey / West Sussex borders have now been assessed for prudent management for butterflies, and that certain areas such as Bookham Common have their tree management especially arranged around the conservation of Purple Emperors so that 'damage limitation' to habitats of butterflies is respected and 'master oaks' are conserved. Willmott saved Turkey Oaks from being felled on one NT site since these were 'master oaks' around which the butterflies establish their territories. He has been instrumental in working with various bodies in managing woodlands for the Purple Emperor and other woodland butterfly species, such as the Brown and Purple Hairstreak. In another NT woodland, volunteers from the then BBCS helped to take out Sycamore (*Acer pseudoplatanus*) which was crowding out the woodland canopy. The key to success is making the various conservation bodies aware of the requirements of the butterfly, otherwise some sites become crowded out and are lost by being overgrown.

The woodland edge and the roadside/trackside edge are also important for the conservation of the Purple Emperor as well as for a host of other butterflies, and should be well managed. There should there be good gaps in the woodland canopy so that the butterfly can use the sunny woodland floor. But there also needs to be bare patches of ground on which the sallow seedlings can germinate. This structural patchiness of the woodland can come about by natural means, such as disease of trees, and from storm damage to trees creating gaps, but is more likely today to be brought about by people managing the wood in a traditional manner.

The cutting of woods in blocks of an acre or two is, however, not always economic sense. Where woodlands remain un-managed for forestry priorities there are always great opportunities for people who just like to encourage wildlife. There are now several land-owners in south-east England who have significantly large private woodlands of about 40-80 ha or more, which they keep simply butterflies. Ride widening not only lets in more light to the

Habitat management

benefit of the butterflies, but makes otherwise waterlogged rides more passable - a necessary forestry management consideration. In East Sussex there is at least one woodland in a classic locality which has been planted with oak and sallow especially for the butterfly, and which has already attracted Purple emperors. Planting groups of oaks is a necessary consideration for the long term conservation of the Purple emperor.

Conservation of the Purple Emperor depends on sympathetic management of its habitat. It is interesting to reflect on how abundant the insect would have been long ago in the then much wooded countryside - secondary woodland - and to picture the old collectors deep in a parcel of woodland just like the Aurelians resting after a good morning's collecting along some woodland ride such as in the New Forest, or some Sussex woodland - with just a glint of purple in the sun. Willmott (1987) provides photographs of some good Purple Emperor sites. His Utopia for the Purple Emperor is as follows, for those interested in following his 35 years of experience with this woodland butterfly: (i) Deciduous woodland: If not entirely, then substantial blocks within commercial coniferous plantations; (ii) Widespread growth of *Salix* sp. (especially *Salix caprea*); (iii) Condensed breeding areas. Approximately 25 m x 50 m minimum (one such area per 100 acres?); (iv) 40-50 mature *Salix caprea* minimum each area; (v) 65-70% Salix sp of female sex (this often occurs naturally); (vi) North-South running rides; (vii) Width of rides - 12-16 m; (viii) Metalled roads - East-West running (verges of such roads minimum of 5 m width); (ix) Habitat with gradient, reaching 200 m + above sea level territories); (x) 'Canopy Gap' territories, near edge of woodland; (xi) 'Canopy Gap' territory - approx size 35m x 25 m (measured on ground); (xii) Perching trees with southerly aspect. (20-24 m high preferably English Oak); (xiii) Sympathetic owners (e.g. Country Trust, National Trust whose management plans take into account the requirements of *Apatura iris*, thus: (a) Designation of priority areas for territories and condensed breeding areas; (b) The maintenance of these; (c) Future planning - opening up of new rides and planting of young Salix sp in specially prepared area (Reproduced with permission of Ken Willmott, from Willmott, 1987.)

Grazing is a necessary tool in the conservation of many grassland butterflies. In coastal areas and on downlands where rabbits fluctuate in numbers the variable grazing regime provides a different range of tall and short sward heights for butterflies. The Butterflies Under Threat Team (BUTT, 1986) had already established that butterflies require various different sward heights in their habitats. Marbled Whites, Meadow Browns and various skippers like the grass long, others prefer it short. Work carried out by Linda and Andrew Barker of the Hampshire Branch of Butterfly Conservation on butterfly populations on Magdalen Hill Down (calcareous grassland) indicated that winter sheep grazing is ideal at regulating turf height suitable for the Brown Argus, as well as maintaining a range of turf heights suitable for other species (Barker & Barker, 1993). Their preliminary results indicated that a certain amount of rabbit activity was good at providing ovipositing sites, but too much made the site unsuitable.

The Grayling, which in Yorkshire is on the verge of extinction, has suffered from the neglect of some sites which have grown a tall sward in consequence. It prefers short turf, which is equally lost when rabbits are killed off by myxomatosis. Sutton & Beaumont (1989) drew attention to the loss of the Grayling from some habitats because of the cessation of grazing, but remarked that the Marbled White has increased its populations on the long grass at the Grayling's expense.(2) In Quarry sites in Yorkshire butterflies have exploited the short turf created on the quarry floor by nothing less than their principal instrument of conservation management, the bulldozer.

In the agricultural areas covering much of the downlands and Chilterns of southern England sites for butterflies are increasingly restricted. Happily research by Feber & Smith (1993) showed that extending the margins of arable fields helped to boost the numbers of butterflies using them, and that the trend amongst farmers to utilize the field margins for wildlife was having a very positive effect.

A classic site for downland butterflies in southern England is Old Winchester Hill NNR which supports numerous species. It has been the subject of various experimental management regimes over the years, but it was in 1981 that it was used for the re-establishment of the scarce Adonis Blue (Thomas, 1987, cf. Oates, 1993a). This species declined rapidly from the mid-1950s and was lost from a quarter of the nature reserves where it was previously known (Thomas, 1984). Thomas showed that one third of these sites were lost because of 'improvement' of ancient downland.

The ecological factors useful for the conservation of this species, discovered by Thomas, were (i) the butterfly restricts egg-laying to areas with short patches of turf, ii) adults do not wander far and leave their colonies only when there are unusually high numbers, and iii) that extinction of colonies occurred 2-3 years after cessation of grazing. Old Winchester Hill was a suitable place to try to re-establish the Adonis Blue since it had become extinct there in the mid-1950s: intensive grazing, using sheep, was then carried out in 16 compartments ready for the re-establishment. In the spring of 1981 65 adults were released onto the site, which resulted in about 4,200 adults in 1982. The population did very well despite two cool summers, and just about survived there with 150 adults seen in the spring of 1985. Writing in 1987, Thomas reported that the colony had survived 14 generations since re-establishment. Rotational management of the compartments ensured that the foodplant Horseshoe Vetch, *Hippocrepis comosa*, was available at least in some compartments. Where it was not available to the Adonis Blues, some of the other butterflies of the downs prospered. Thomas (1987) concluded that the Adonis Blue could be successfully conserved in Britain if sites are adequately

Habitat management

grazed, but that deliberate introductions are necessary in isolated areas.

MANAGEMENT UNDER DIVERSE OWNERSHIPS

Butterflies are increasingly found in parts of the countryside that have passed into government hands (such as the MOD or FC) or are owned by independent conservation bodies either backed by government (such as the NT) or who have no specific government acts to support their conservation efforts (such as the RSNC or Woodland Trust reserves).
 A recent survey of a large segment of the southern England including Buckinghamshire, Oxfordshire, Berkshire, Wiltshire, Hampshire, Isle of Wight and Dorset, carried out by Martin Warren, showed that the major landowner of key butterfly colonies (with nearly 10%) was the National Trust; the next largest was the Forestry Commission with 8%, followed by the Ministry of Defence (about 7%) and various Public Authorities (7%) (Warren, 1993a).
 Considering that the greatest richness of butterflies is in the south of England, the area chosen by Warren was appropriate and is fairly representative of other parts of southern England. It typically shows the division of the land into the various bodies, a trend which can be extrapolated nationwide. So how do butterflies fare under the different owners and managers?
 The Ministry of Defence with their off-limit lands are doing superbly, getting top rating across their many properties with the most species-rich habitats. Denying people access is always good for promoting butterflies. A look at the MOD's publication *Sanctuary* will illustrate their combined effort at conserving butterflies whilst making some considerations in terms of habitat management for them.
 Such has been the piecemeal carve-up of the countryside by all sorts of interested parties that it is alarming to learn that less than about 10% of the total area of such counties as Berkshire, Buckinghamshire and Oxfordshire is now suitable for butterflies as semi-natural vegetation; almost 90% being unsuitable for butterflies as improved farmland or roads and houses (Steel and Steel, 1985). How much of this can be said of other counties in Southern England? What of the fate of butterflies in these remaining refuges, and, if conservation policies are present, is there consistency of policy, or any policy at all?
 The National Trust is one of the largest conservation organizations in Britain. With its large holdings it is inevitable that it supports some large populations of butterflies; and this is particularly the case with its special

Habitat management

acquisition of coastal habitats. The NT therefore carries the responsibility for conserving some of our most prestigious butterflies, including many species of blue typical of chalk grassland, such as the Chalkhill or Adonis Blue. It even helps to conserve the Large Blue in a very significant manner, but keeps very quiet about it. They are also the largest single owner of Heath Fritillary sites in England, which are on Exmoor.

The long-term strategy of the NT to acquire as much of the best parts of the coastline of Britain under the Operation Neptune scheme has been relatively successful over the years. Ecologically, good coastal sites are naturally good for butterflies so it is not surprising that those areas conserved for the NT have profited well for butterflies such as the Grayling which enjoy these grassy headlands.

The good thing about the NT and the way they acknowledge responsibility for habitats and wildlife is that they have formulated comprehensive management reports of all their sites. This includes ecological assessment of the sites around which proper management plans are published, and later reviewed. These 'internal' reports are used for the everyday management of all its sites, and as a reservoir of information for protection of sites at public inquiries. They are not public documents, but as far as I can ascertain from the few reports I have seen, they are the most comprehensive one could expect on any habitat.

The NT have only recently concluded a nationwide study of their ecologically important sites. Although beetles might have been found more important on some ancient woodland sites and parks, and might have eclipsed butterflies for a time, there remains a huge extent of NT land that is full of the commoner species of butterflies. The future of many butterflies is therefore secured on these extensive lands, whether or not the butterfly species have been collated there.

The NT has an inbuilt indirect conservation strategy in favour of butterflies. It buys or is given properties which often come with lots of countryside. So, although neither butterflies nor their habitats have been targeted as specific acquisitions, the country estates come complete with their package of invertebrates including butterflies. The NT are in the business of protecting places of natural beauty for the enjoyment of the public. Once in NT ownership or management butterflies are likely to fare much better than in most other statutory or voluntary body's hands. NT land is virtually inalienable for ever, which must be good news for all butterflies, and butterfly-lovers; it is only under attack from aerial pollution and the effects of global warming or persistent government acquisition.

As befitting a conservation body at the leading edge of the field, the NT have published and continued to up-date, the policy statement *'The National*

Habitat management

Trust and nature conservation' (NT,1980) in which references to butterflies are often made. Wicken Fen is usually highlighted as the site that illustrates how habitats have to be managed in order to protect a particular part of the countryside, in this case a typical part of the fenland which has as its gem, the Swallowtail. But there are many other segments of the British countryside that are offered conservation and protection of habitat through the NT. These include many chalkland sites on the Sussex or Surrey Downs (such as Box Hill, Surrey) or the Chilterns where butterflies still fly in great numbers. For a review of the National Trust's grazing strategies using old cattle breeds, ponies, sheep and bullocks, readers are referred to Oates (1993a,b., Oates, 1995) Managing the wild flowers, the foodplants of the butterflies, the NT are able to nurture many of the common and less common blues as well as the browns and skippers.

In April 1995, the National Trust for Scotland almost doubled its land holding in Scotland when it bought the 120-square mile Mar Lodge Estate for £5.5 million, adjacent to the Balmoral Estate, with money from the National Lottery and from a private charity (£4 million). The site is regarded as the most spectacular remaining wilderness in Europe and is now safe in the hands of an organisation whose main interest is conservation. The RSPB and WWF were, at one time, bidders. This is the first time that National Lottery money has been used for nature conservation. (*Independent*, 27 April, 1995). The butterflies which live within this huge estate will no doubt be conserved sympathetically.

On the question of whether to conserve single species of butterfly or go for whole habitat conservation there is a growing momentum sensed within NT for the latter (Oates, pers. comm. 1994). We are, though, fully committed to single species conservation in Britain whether we like it or not, something that the managers in the NCC bequeathed us. All recent conservation work on NT properties has been done with reference to their strategy document (NT, 1989). The very latest consideration for butterflies is now embodied in their Countryside Policy Review published early in 1994 which argues in an interesting way that single species conservation is too narrow a focus for management of the countryside (Oates, pers comm. 1994). Serious considerations might now be expected from the Second Policy Review whose views are a little too finely focused. It would be an opportune time to see bodies representing different persuasions coming together on the conservation strategy of mutually-important habitats for the good of those key habitats and flora and fauna. In the meantime Butterfly Conservation is taking the lead with its 1995 commitment to reverse the decline in Britain's butterflies by targeting 25 threatened species for conservation.

Habitat management

Notes

1. Bernwood Forest (400 ha) was declared a Forest Nature Reserve on 21 April 1981 when its owners, the Forestry Commission, signed a 20 year agreement with the NCC. The aim of the agreement was 'to grow and harvest marketable timber with minimum expenditure and to maintain the floristic and structural diversity in the woodland for the benefit of butterflies and moths'. It is through the pursuit of the Black Hairstreak that the full entomological potential of Bernwood became known, reaching at its peak some 42 species of butterfly including Purple Emperors and most fritillaries; the NCC (1983) recorded 39 species of butterfly including nine rare species. Bernwood is particularly rich in rampant Blackthorn hedges and these have been left deliberately to encourage Black Hairstreaks, especially along rides giving a warm south-facing aspect. The research done on the Black Hairstreak at Bernwood typifies the value of autecological studies.

2. The chapter on conservation of sites and species in Yorkshire as published by these authors is exemplary and unsurpassed in other county books on butterflies.

Habitat management

Habitat management

CHAPTER NINE
Threats to butterflies

INTRODUCTION

This chapter deals with the threats to butterflies, mostly through disruption of their habitats. The aim is to highlight the threats so that they can be recognized and countered in the future. Although this book is about conservation, the threats to butterflies have to be recognized before conservation can take place. It is through the recognition of threats (which often comes too late to save butterflies) that action can be implemented. A range of examples have been included where people have been successful at stopping the threat from damaging habitats and their butterflies. Details are given of a few examples where butterflies and their habitats have been under pressure on Sites of Special Scientific Interest where they are meant to be protected.

Threats to butterflies

It is thought in some quarters that 'factors such as physical or genetic isolation, butterfly collectors, pesticides, air pollution or a changing climate' are all regarded as an out-dated rationale which was used to explain the decline in butterfly populations in the 1950s and 1960s (Erhardt & Thomas, 1991, Thomas, 1984). This is rather a sweeping statement and should not be accepted as the whole truth. Obviously habitat loss has been the principal enemy to butterflies and their habitats, as any view across lowland Britain will indicate, but the other factors mentioned above should not be so cursorily dismissed as if they did not, and do not still, add to butterfly extinctions just about everywhere. It is too easy to dismiss them without the evidence.

The general decline in the number of species of butterfly is worrying. In Yorkshire, which shows some trends typical of many counties in Britain, there has been a considerable net contraction and extinction of butterfly species since 1900. Altogether some eight former resident species have disappeared and six species are reduced to a handful of sites (Sutton & Beaumont, 1989). The decline in butterflies is also seen in southern England, where it is even more worrying since this is where the greatest richness of species occurs in Britain. In his comprehensive survey of the state of butterflies in six counties in southern-central England, Warren recorded the extinction of 384 colonies (out of 2248) over the previous 40 years up to 1987 (Warren, 1993a,b). Woodland butterflies such as the Pearl-bordered and Small Pearl-bordered Fritillary, the High Brown Fritillary and the Wood White had declined through the cessation of coppicing, which was also thought to have been responsible for the demise of the Chequered Skipper in England. On the other hand both the Adonis Blue and Silver-spotted Skipper ceased to decline through the 1980s, probably because rabbits were recovering after myxomatosis so that the short swards the butterfly required became available again. However, populations of Silver-studded Blue continued to decline after 1980. This monitoring of the health of butterflies in this prime part of England threw up some invaluable information on the state of butterflies, but it has not been continued. This is unfortunate because understanding the threats to butterfly populations, particularly the way that woodlands are neglected, and the way that coniferous forests grow up and occlude the vital sunlight and dappled light that butterflies need, is half the battle.

The distinction is made here between habitat loss and habitat degradation. Habitat loss is the entire loss of a locality or habitat where a butterfly once lived. Habitat degradation is the interference of that habitat or locality, which makes it less rich than it was, and may involve partial habitat loss. All sorts of human activities contrive to decrease the ecological appeal of the site.

Threats to butterflies

LOSS OF THE BEST BUTTERFLY SANCTUARIES

The basis of nature conservation in the UK is the protection of particularly good habitats and their flora and fauna as a series of sites called Sites of Special Scientific Interest, more widely known as SSSIs. The term is a good reference point for naturalists, conservationists and the general public, since it is used regularly in the national press, and most people who are now interested in the living world know all about SSSIs. The idea of SSSIs is that the habitats, flora and fauna found within them represent the cream of the cream of British wildlife, of which butterflies form a considerable and colourful part.

There are other perspectives to SSSIs than the biological one, since some are important geologically. In some cases the nature of the geology is such that it produces an ideal habitat for butterflies; such is the case along much of the south coast of the Isle of Wight where the regular slumping of the cliff and soil provides habitat for the Glanville Fritillary. English Nature themselves define SSSIs as 'the best examples of our natural heritage of wildlife habitats, geological features and landforms' and they notify them 'because of their plants, animals, or geological or physiographical features'.

The first SSSIs were set up under provision of the National Parks and Access to the Countryside Act 1949. With the Wildlife and Countryside Act 1981 came the statutory requirement of the Nature Conservancy Council (NCC) to notify new SSSIs. Now it is the turn of English Nature to notify or denotify SSSIs and to keep them regularly reviewed.

As of 30 September 1993 there were 3,749 SSSIs in England covering 861,341 ha covering about 6% of England's total area, and in the ownership of or occupied by about 18,000 people (EN, 1993b). Most SSSIs are in the ownership of these people and 40% of the SSSIs are owned by or managed by public bodies such as the Forestry Commission, Ministry of Defence and Crown Estate Commissioners, or by groups in the voluntary conservation sector.

Under the 1949 Act the SSSIs had to be notified to local planning authorities so that consideration could be given to their conservation in the planning process. Under the 1981 Act an SSSI has to be formally notified to the owners and occupiers of the land, and to the Secretary of State for the Environment as well as the local planning authority. Until owners and occupiers have been notified of the presence of SSSIs on their land (some would not always know they had them) the regulations of the 1981 Act do not apply. At the beginning of 1994, English Nature had largely completed notification, renotification and denotification of SSSIs in England.

Diverse ownership has been a stumbling block to butterfly conservation. A contributory factor in the loss and damage to some SSSIs (and thus their wildlife including butterflies) has been the relative secrecy of actually where the

Threats to butterflies

SSSIs were. People might have heard about SSSIs but still do not know where their nearest one is. English Nature, and the NCC before them, have not been at liberty to disclose to members of the public the location of SSSIs since this would have breached the confidentiality of the owners of the land, some of whom would not take kindly to their land (over which there might not be any legal right of access) being walked over by all and sundry, even *bone fide* naturalists. Some large estates (some examples are in the West County) enclose enormous SSSIs, yet historically have never had any public rights of way (see, for example, Shoard, 1980,1987).

I have made extensive enquiries to find examples of butterfly-rich SSSIs that have been lost through development. I enquired of the 47 Wildlife Trusts but was informed of only a handful of locations. This may be an indication of a poor response rate (see Chapter 10 for SSSIs saved). It may be that butterflies have fared rather better than expected on SSSIs, although the annual loss and damaging procedures that affect SSSIs must have a direct impact on butterfly populations even if they are not recorded. One also has to consider that there is still a lot of under-reporting of key butterfly sites within the Wildlife Trusts and that some Trusts are unaware of sites or of their significance in a national context.

Perhaps the most recent examples of butterflies lost from SSSIs are those relating to the Marsh Fritillary in Glamorgan. The case at the Parc Slip West site goes back to 1989 and involved a Public Inquiry relating to SSSI issues and implications of the (EC) Habitats Directive (see chapter 10). The site holds the fourth largest colony in West Glamorgan, the tenth in Glamorgan as a whole, and the largest colony in West Glamorgan outside Gower (Owen Lewis letter to Martin Warren, 1993; from the Conservation Office of the South Wales Branch). The butterflies and their habitats on the development site were lost. The verdict on the case (published in September 1993) held that open cast mining could go ahead on the site despite the presence of Marsh Fritillary habitats. The Habitats Directive was signed in June 1994 but cannot be made retrospective. As it was not in force during the duration of the Public Inquiry until the Inspector made his decision, it did not legally apply. However, there is much in EC law that encourages legislators and those who judge the law to abide by 'the spirit' of the various laws and directives. That the Habitats Directive was in effect in place waiting to be signed is a missed opportunity which has not advanced nature conservation in Britain.

The attempts at conservation of the Park Slip West have not been helped either from the administration point of view, or from the environmental survey point of view, as the boundary between West Glamorgan and Mid-Glamorgan goes through the site. All the nature conservation interest of the site is in West Glamorgan. It was not until late 1989 when a planning application was received that attention was focused on the site for the first time. It had never been assessed ecologically before. A quarter of the area was an industrial site with building materials, and another quarter of the site was an old open-cast site last worked in 1955-58 (at least pre-1960). After considering the application for re-working the site, West Glamorgan approved the project, but Mid-Glamorgan turned it down, thus precipitating a Public Inquiry. In order to find out more about the precise distribution of the butterfly in the area a Phase I project (1) was undertaken in the West Glamorgan area affected, but this has been supplemented by further Phase 1 fieldwork on a field by field basis by ecologists - a task which is due to be completed by 1996 but will probably take until the year 2000. In the meantime two wet pasture sites have been designated as SSSIs to protect the butterflies. Butterfly Conservation have been instrumental in promoting awareness of the plight of Marsh Fritillaries in Wales, especially through the work of their conservation officer, Martin Warren who has done a lot of field work on this species in Britain.

There is another site to which Butterfly Conservation have drawn attention to, and this is the Selar Farm Grassland SSSI near West Glamorgan. This case concerns another Marsh Fritillary site in Wales which was lost in late 1994 (Butterfly Conservation, 1994b). The Secretary of State for Wales gave permission for the bulldozers of British Coal Opencast to move in despite much opposition from Butterfly Conservation, the Countryside Council for Wales and the Allerdale District Council. The case did not go to Public Inquiry. The Marsh Fritillary is given special protection under Schedule 2 of the EC Habitats and Species Directive, and is the only native British butterfly to be included. Both the butterfly and its habitat - a uniquely rich mixture of meadow plants which comprise a threatened European habitat - have now been destroyed at this site. There was much discussion about adjacent localities being suitable for the butterflies, and some misguided assumptions that the butterfly could just decamp and fly over to the next suitable habitat. There is much to learn from this failed attempt to conserve a notified SSSI butterfly site, particularly in getting the entomological case across to lay persons including those people who have to make informed judgements about the future of the countryside and its wildlife.

Paul Kirkland, Assistant Conservation Officer of Butterfly Conservation, who wrote a resume of the Selar case (Kirkland, 1995a) highlighted the extraordinary allocation of vast funds to translocate some of the butterflies and habitat for a process which is certainly not proven. In conservation terms the money could have been more effectively and efficiently allocated with a guaranteed return for all parties concerned in the matter. To quote Paul Kirkland 'Only 6 acres of the 17 acre SSSI are to be "translocated", and using comparable costs from other translocations, we estimate that the final cost of

Threats to butterflies

this will be over £0.5 million or nearly £100,000 per acres!!' Butterfly Conservation could have bought over 100 acres of similar, SSSI-quality Marsh Fritillary habitat if they were give the money instead. If the Selar case had gone to Public Inquiry Butterfly Conservation would have argued the case that translocation is not a proven technique.

Indeed there is a precedent case, where proposed developments on the Maryport Harbour SSSI for a marina were refused by the then Secretary of State on the grounds that translocation was an unproved method. Important flowers and butterflies were involved. The site was regarded as unique by EN on account of the combination of diversity and rarity of the species in the SSSI. At issue, on the entomological side, was one of the strongest colonies of the Small Blue in Cumbria. It is a nationally restricted, declining and vulnerable butterfly. It was thought that the loss of the major part of this strong colony would lead to the total extinction of the species from the site, and, bearing in mind that the prospect of a successful translocation programme of moving the butterfly and habitat off site was relatively poor, the Inspector recommended refusal of the development. (Inspector's Report. *Application by Maryport Developments Limited, Senhouse Dock, Maryport Harbour, Maryport, Allerdale, Cumbria.* Annex to the letter of 19 November 1992 from the Secretary of State for the Environment. 14. Inspector's Conclusions. pp. 30-42.

In yet another case which preceded it, the Small Blue was again the focus of attention at a Public Inquiry on the Isle of Wight in 1989, and won the day. The case at hand was a re-alignment of a coastal road over Afton Down SSSI owned by the National Trust. The fact that the new road to be built was smaller than the existing one it was to replace did not sway the Inspector who upheld the inalienability of NT land, as well as the importance of the SSSI as a whole, and the butterfly. New habitat would have been made available for the butterfly.

The fact that the spirit of the EC Habitat Directive was over-ridden at Selar Farm for a protected species, is a landmark case, and it does highlight the vulnerability of notified sites. If there is an overriding need to have a national or regional project, then SSSI designation can unfortunately be overruled. SSSIs are theoretically protected via the National Parks and Access to the Countryside Act and the Wildlife and Countryside Act 1981 and any potential impact should be discussed through the regular planning process.

MEADOWS AND GRASSLANDS

The need to repeat that 97% of Britain's meadows have been destroyed seems a little unnecessary since we have all been bombarded with the statistics from various quarters (cf. Feltwell, 1991). This loss means a simultaneous loss of many butterflies from those meadows. The worry is that we have become immune to the significance of the statistics, un-responsive to the figures that we see in every countryside magazine we pick up. Conserving butterflies in the remaining 3% of Britain's flowery meadows is what many are all working at today. Making do with what we have left only concentrates the mind and makes us more mindful of the importance of butterflies in the British countryside. No wonder people become significantly motivated when a meadow is under threat.

Meadows are not now studied for their own sake, and rarely for butterfly conservation. Government money is being spent on some aspects of meadow ecology but the project aims have to be clear and in line with some government policy or concern (e.g. effects of global warming). For instance grants are made available for studies of changes in vegetation caused by global warming, improvement of field margins for wildlife, transplanting meadows, and invertebrate surveys of meadows, e.g. Old Winchester Hill in Hampshire. These are the sorts of precise aspects of research that English Nature are funding.

Agricultural 'improvement' has probably been the most insidious threat to meadows in Britain. Changing the regime of the old meadow from 'unimproved' to an 'improved' status by applying fertilisers or by drainage, eliminates the diverse flora immediately and eliminates most of the butterflies. Other threats to meadows, and thus to butterflies, have been expanding housing and industrial parks on green field sites.

Road infrastructure has threatened meadows too. In a recent case, that of the old medieval Lugg Meadows in Herefordshire, a Public Inquiry in 1993 concluded that the environmentalists were right after all and that the major road scheme would have inflicted unacceptable damage on the countryside (cf. *Countryside Campaigner*, Autumn 1993, p. 8). The Herefordshire Lugg Meadows are particularly important since they are the largest expanse of common meadows, or Lammas meadows, in Britain. In my 1991 *Meadows, a history and natural history*, I summarized the best Lammas meadows in Britain which remain in Gloucestershire, Huntingdonshire, Oxfordshire, Worcestershire, Hertfordshire and Middlesex, Wiltshire, Somerset, as well as others in Herefordshire. They will all be havens for butterflies.

English Nature is busy promoting the conservation of grasslands and butterflies on Jurassic limestone in the Cotswolds and high quality semi-natural neutral grassland in Worcestershire. There are only about 4000 ha of semi-natural western Jurassic limestone grassland left in England, and the Gloucestershire Cotswolds have 2,500 ha or 60% of the resource, complete with butterflies. £25,000 was given to the Gloucester Wildlife Trust in 1992/3 for identifying areas of nature conservation value, and for management enhancement (EN, 1993e).

Threats to butterflies

Meanwhile the Worcestershire Neutral Grassland Survey has completed its survey of 623 sites of high quality semi-natural neutral grassland comprising 1,273 ha, and found somewhat alarmingly that since 1980 37% of fields had been destroyed and have lost all nature conservation interest, largely because of agricultural intensification ('improvement'). Another 28% had lost some of their interest, more than 50% because of management neglect. Thus Worcestershire has lost 65% of its neutral grassland since 1980, and it now has only 787 ha of this resource, with butterflies. The Worcestershire Nature Conservation Trust Ltd received £30,000 from English Nature during 1992/3 for work on this survey (EN, 1993e).

HEATHLANDS

Heathlands have suffered a similar catastrophic habitat loss to meadows during the same period and this has impacted negatively on heathland-dependent butterflies. The Silver-studded Blue - the lepidopterists' equivalent of the Dartford warbler - is a typical heathland species which breeds on a variety of heath-loving plant species, such as Heather (*Calluna vulgaris*), Gorse (*Ulex europeaus*), Bird's foot trefoil (*Lotus corniculatus*) and Rockroses (*Cistus* sp.) - and it too has had mixed fortunes.

The plight of this species in Suffolk stimulated the Suffolk Trust for Nature Conservation to look carefully at the remaining distribution, ecology and mobility of the species in the county (Ravenscroft, 1986). The butterfly has now been lost from the Brecklands and the extreme north-east of Suffolk, and in its remaining coastal areas known as the 'Sandlings' was still being lost in the late 1980s. Reasons for this decline in heathland habitat have been (i) ploughing (ii) afforestation of heathlands, (iii) neglect (iv) loss to development, all of these contributing to (v) isolation and fragmentation of suitable habitats so that distances between populations are too great for the insect to maintain connections.

The report on the Silver-studded Blue revealed there were just nine colonies left in Suffolk, the largest being one of an estimated 20,000 individuals (over the flight period) at Martlesham Heath (which was due to be partially destroyed for a housing development). The other colonies were below 2,500 individuals strong and five were less than 1,000 butterflies. As for mobility, the species was found to be fairly sedentary, females being slightly more adventurous than males. Thus the importance of fragmentation of habitats is a truly isolating factor in the life of the butterfly. The presence of Bell Heather (*Erica cinerea*) was important since females feed almost exclusively on this. And presence of ants was very important also since these tended the larvae, three species being involved: *Lasius niger, Lasius alienus* and *Lasius flavus*; the greater the number of ant nests on a heath the more abundant is this butterfly (Ravenscroft, 1988).

In 1984 Sainsbury plc built a supermarket on the site of a colony of Silver-studded Blues near Ipswich, Suffolk. There was local and national furore over the loss of the butterfly habitat, the result of which was a much publicized attempt to transplant the butterfly habitat piece by piece from the proposed site of the supermarket to a receptor site. This was done with much publicity enjoyed by Sainsbury's in the local press. The result was in fact a failure for the butterflies which ceased to exist on the new site, although the vegetation and ants survived (Ravenscroft, 1988).

What can be learnt from this publicity event? The supermarket gained more than the butterfly conservation world. The butterfly certainly did not gain anything except local extinction. The media event certainly stimulated local conservationists to find out more about this butterfly, and to commission autecological research. It brought public awareness to focus on one species, and served to highlight the plight of Suffolk heathlands. Visitors to this Suffolk Sainsbury's are reminded by a logo of the butterfly, not the real thing. Transplanting vegetation is far easier than transplanting the butterfly. The money spent on this publicity stunt would have been perhaps better spent backing some autecological study: 'Unfortunately the wholesale removal of the Silver-studded Blue site near Ipswich was a total failure as many had predicted it would be, and emphasises one factor which must always be remembered: it is vital that Branches and members monitor most carefully any notifications in the local press concerning applications for outline planning consent' (*News* of the BBCS, No 36, Spring 1986, p.11, Chairman's Report.)

The plight of butterflies on Suffolk's heathlands again made the local newspaper headlines in 1985 when Outline Planning Permission was given for 122 homes to be built by Bradford Property Trust developers on a former locality of Silver-studded Blues, Martlesham Heath (*East Anglia Daily* Times 27 June 1985 and 3 July 1985). This heathland offered refuge for the largest population of Silver-studded Blues in Suffolk and half the heathland was proposed to be built over.

This was a case of Outline Planning Permission being granted before the significance of the site was known from the Suffolk Butterfly Survey. Remedial methods would have been second best, even bound to lead to the significant loss of butterfly populations. Arrangements were made with the NCC and Suffolk Trust for Nature Conservation to ensure that the rest of the butterfly populations were fenced off from the development. Another proposal by the Suffolk Coastal Council planners to transfer the top soil and vegetation to another site - a local runway - was made but did not materialise. All the remaining Silver-studded Blues in Suffolk are now within the Suffolk Coastal Council area (News of the

Threats to butterflies

BBCS, No.35, Autumn 1985, p. 18) so there should now be a concerted effort within the council to conserve these butterflies for the future.

OTHER IMPACTS

There are two major environmental impacts which have removed or damaged suites of protected sites, and these are the M25 developments and the Channel Tunnel development. The M25 has impacted upon chalk escarpments of the North Downs in Surrey and the Wealden interface, whilst the Channel Tunnel has impinged upon the North Downs, coastal cliffs, stream margins, woodlands, meadows and agricultural land.

The Channel Tunnel development to the north of Folkestone (Kent) has taken a significant section of countryside - fields, woods and meadows - out of the ecosystem for ever, including the whole of Biggins Wood. It has also negatively impacted the Shakespeare Cliffs SSSI, the Folkestone-Etchinghill Escarpment SSSI and the Folkestone Warren SSSI, which were part of a continuum of coastal habitat which included the White Cliffs of Dover - and where butterflies flew. Scientists were allowed onto various parts of the site, to help in ameliorating the damage and finding out what was there; it was reported by Anonymous (1989) that 38,000 invertebrate animals had already been collected on the site, and that at least two butterfly transects were set up between Castle Hill and Sugar Loaf hill, both interesting ecologically. Reports of the transects were not published, but the key species of the area were the Brown Argus, Speckled Wood, Marbled White, Common Blue and the Adonis Blue.

The loss of Britain's hedgerows during the 1960-80s is well documented elsewhere and can only have deprived butterflies of plenty of habitat. Suffolk, again, has well accounted for the plight of its butterflies in that part of East Anglia where hedgerow loss has been more significant than many other places. The largest colony of Suffolk's White-letter Hairstreak was threatened in 1986 with the 'piecemeal' development of the Bixley Farm Development (*East Anglia Daily Times*, Friday 5 December 1986). Outline Planning Permission was sought for a development that impacted negatively on a particular elm hedge which supported Suffolk's largest population of White-letter Hairstreaks in a buffer zone between Rushmere Heath and Bixley Farm. The Ipswich Area Conservation Group sought to preserve the hedge and the butterflies and to maintain the hedge in the buffer zone.

AGRO-CHEMICALS

The proliferation of organic chemicals in the late 1940s and 1950s and their hurried and widespread use in seed dressings and roadside treatments caused some alarm within the ranks of the JCCBI if what is recorded in the Minutes is anything to go by. No-one seemed to have any information about possible effects on wildlife. The JCCBI recommended that birds, trout and honeybees be tested in the laboratory, and that this was really the job of the manufacturer, rather than the ministry. Whilst all this was happening spraying was being carried out on along the roads and lanes of England. The Nature Conservancy, which had convened a meeting in September 1951 on the Effects on Wildlife of Toxic Chemicals, drew attention to the lack of information on the effect of toxic chemicals (Minutes of 8 January, 1958). As they say, the jury was out on this one; it still is.

If one has to have evidence (qualitative I am afraid) that roadside spraying kills insects, then it is recorded in the minutes of the JCCBI of 18 October 1960 that Dr Massee recorded the extinction of a rare (and un-named) myrid from the roadside at Charing (Ashford, Kent) through spraying.

Read all the butterfly books and papers (especially those of Thomas and Morris) and you will get the clear and repetitious message (the same as with over-collecting) that upon inspection of all available evidence there is nothing to show that chemicals have caused the demise of butterflies. There is, however, an element of scientific inaccuracy. We have absolutely no idea how much butterflies have been impacted by chemicals. They do not all conveniently go back to a hive to die, like honeybees after a dose of insecticide (where quantitative evidence is then possible), so it is impossible to conduct any experiment in the field to determine the effects of chemicals. It is a pity that butterflies are not mammals or birds which are simple to analyse.

Agricultural chemicals, such as pesticides, herbicides and fertilizers are extremely difficult to monitor once they have been sprayed. Where they go to in the environment has not been easy to sort out (Davidson & Lloyd, 1977). It is easy to sit back and theorize that the bad ones are leached through the soil to the local water course, playing ecological havoc all the way to the sea. In effect a lot of chemicals today are no longer active once they touch the soil so their persistence in the environment (a useful monitor) can, theoretically, be measured. In the absence of work on butterflies, we have to look to the work done on aquatic plants and fish. Some contrasting and surprising things have been found, such as ponds completely clear of persistent chemicals even though surrounding land has been heavily sprayed, and, in great contrast, fish with relatively high levels of the very same chemicals. Perhaps the most insidious of all chemicals, however, are herbicides which eliminate many of the foodplants

Threats to butterflies

of larvae from marginal areas.

Butterflies are perhaps most at risk when they are on the wing, particularly when nectar-feeding. The risk increases dramatically when the nectar they are seeking comes from a crop species which is being sprayed, such as Oilseed rape, *Brassica napa*. The large fields of lucerne and clover that used to be grown widely in southern Britain earlier this century, and which were a useful nectar source of butterflies, have now been replaced by even large expanses of newer crops, such as Oilseed rape.

Few ecologists would attest that butterflies have never been affected by chemical sprays. The question is how much have they been affected. We would all like to know. If they did not die on contact, but could not breed as a result, we would not be the wiser. Birds and mammals have been so affected (although the herbicides and insecticides have not been targeted upon their physiology), so why not butterflies since butterflies are insects, and insecticides are targeted at insects. Butterflies have softer bodies than a lot of other insects such as beetles, so the chances of absorption of chemicals through cuticle and wing membrane are even greater, especially when beetle pests are the ones that are often targeted in the first place.

Robert Pyle (1981) in his concise overview of conservation of British butterflies does not come down on one side or the other on the controversial topic of the possible 'role of insecticides in the alleged decline of British butterflies'. He relies on the only experimental work done, that by Dr Frank Moriarty at Monks Wood Experimental Station (Moriarty, 1969). Moriarty did not do a comprehensive study of all the pesticides available (he tested only DDT and Dieldrin, both persistent organochlorine insecticides) and turned up only a very small bit of evidence that they *might* have an effect on butterflies, but concluded that his work did not suggest that insecticides caused the alleged decline in the numbers of butterflies. It is fortunate, nevertheless that in recent years there has been a reduction in the use of pesticides and herbicides because of the increasing costs of agro-chemicals and the use of integrated pest control.

ACID RAIN

There appears to be no evidence that acid rain has decreased populations of butterflies or made them extinct. In any case, such evidence would be difficult to collect and quantify. From a conservation point of view it would be very useful to know how acid rain is affecting populations of butterflies, so that attempts could then be made to protect butterflies in areas less affected by acid rain. Now that maps of acid rain deposition for England, Scotland and Wales, have been published, we have the basis for conservation research which could be applied to certain butterflies under threat.

Both English Nature and Plantlife published maps of acid rain figures in England, Scotland and Wales (Bisset & Farmer, 1993, Plantlife, 1993), English Nature's maps being the most comprehensive. Data are not provided for each of the SSSIs in Scotland, as it is for each of those in England, in the English Nature publication, and at present no such detailed analysis for SSSIs in Scotland is published. Regions to be hardest hit with acid rain in the future, even if there is a projected reduction in current emissions by 60% (desired goal), are South Wales, North Wales, East Midlands, the Lake District, the Pennines and the Argyll and Perthshire mountains. There are an awful lot of butterflies in those areas, and it is particularly worrying that as there is a trend for contraction of butterflies westwards, the highest acid rain is recorded for these areas as well. Even with a reduction by 80% of emissions which the UK government could do, all these areas would still receive some acid rain, with the hardest hit area being around Manchester and Liverpool.

If the butterflies themselves (and their early stages) are not harmed, killed, or stressed by other factors, then their habitats, particularly their foodplants and cover may be so harmed by acid rain as to threaten the well-being of the species. There are those who might argue that butterflies do just as well outside SSSIs and they do not have to be in an SSSI or even an NNR to be adequately conserved; it is then just as well to consult the maps of acid rain which shows that acid rain is no respector of SSSIs, National Nature Reserves (NNRs), Local Nature Reserves (LNRs), Forest Nature Reserves (FNRs) or plain old countryside where butterflies usually thrive.

Six hundred biological SSSIs representing 16% of the total SSSIs in England were considered at risk from continued acidification by Bisset & Farmer, 1993. If emissions are reduced by 50% by the year 2005, according to the EC Large Combustion Plant Directive, then it is predicted that only 7% of the SSSIs of England will be affected by acid rain. This reduction in affected sites, which represents more than a 50% drop, is to be welcomed. However, the commitment to reduce emissions and the actual results to be seen by 2005 may not reach their target. Hopefully there will be a reduction in the overall impact of acid rain on butterflies and their habitats over this period and beyond. The conclusions of Bisset & Farmer are that even if the target of a 50% reduction in emissions is achieved then this still will not be enough to eliminate acid precipitation from 250 SSSIs in England. The worst affected areas are located throughout England, especially in upland areas of the north-west and south-west and also in the dry acidic soils of southern and eastern England.

The potential environmental impact of acid rain on butterflies is very real. The weakening and stressing of the foodplants of both the larva and adult butterfly could result in butterflies being unable to finish their life cycles. At

Threats to butterflies

worst the foodplant would disappear entirely. As many lower plants (especially mosses and lichens) are under threat from acid rain, that some upland butterflies may be denied shelter and overwintering sites, and be more exposed to predation. The Plantlife team showed that already 100 species of plant (mainly mosses and lichens) had already suffered serious population declines over the last 100 years or more, as a result of acid rain (Plantlife, 1993). The follow-on impact onto insects, and particularly butterflies is therefore a step more predictable.

Note

1. A Phase 1 report sets basic standards for mapping any site according to well laid down procedures such as the use of various prescribed colours for highlighting different environmental characteristics and the use of particular symbols drawn on a map.

CHAPTER TEN
Butterflies and the Law

'habitat is being lost at such a rate that the options for site conservation are diminishing fast ... the outlook is catastrophic' Alan Stubbs, 1981b

'When I was young 46 species were found near my home. I daren't tell you how few are left. Unless we make a gigantic effort to cooperate, tomorrow we may find Red Admirals and Peacocks on the verge of extinction'. Miriam Rothschild, in Butterfly Conservation's Press Release of 21 February 1995.

LAWS FOR BUTTERFLY CONSERVATION

In conserving butterflies, there have to be two considerations, conservation of individual species and conservation of the butterfly habitat. For some butterflies (e.g. the larger fritillaries) the breeding area (i.e. where the larvae

Butterflies and the Law

will feed, and where their larval foodplants are found) may be a completely different place from the type of habitat required by the butterfly. Thus in the conservation of one species, legislation might be required to conserve two types of habitat, or a mosaic of inter-connected habitats. Both individual species and habitats have equal right to be protected in the legal framework.

UK law has to take due consideration of International and European law as well as its own. The UK has signed various international laws and agreements and thus has obligations and commitments which affect butterflies and their habitats. As a general rule, laws of an international nature have supremacy over community law (i.e. emanating from the European Union (EU)), and EU laws have supremacy over UK law (unless the UK begs to differ and it is written down in the statutes that it wishes too). This means that in cases of dispute, international law over-rides EU law, and EU law over-rides UK law.

Laws relating to butterflies are therefore treated here first from an international level then a European level and finally at a national level. One has to consider also the implications of agreements such as the General Agreement on Tariffs and Trade (GATT), since this affects the state of the UK countryside just as much as elsewhere in Europe, and indirectly the fortunes of butterflies and their habitats. The effects of GATT will probably have as much environmental impact as the General Agricultural Policy (GAP) of the 1970s. It is naive to think that it will not. In the meantime Britain's butterflies are at the mercy of the management of the land, as ever, as controlled from Westminster, Washington or Brussels.

INTERNATIONAL LAWS

There are four international conventions that protect wildlife signed by the UK government, but few list butterflies for conservation. These include the World Heritage Convention, CITES, the Bonn Convention for migratory species, and RAMSAR agreement.

In 1972 the UK signed the 'World Heritage Convention' which is properly called The Convention concerning the Protection of the World Cultural and Natural Heritage'. This seeks the establishment of 'natural' world heritage sites, and in doing so serves to protect flora, fauna and habitats within those sites.

In 1979 the UK signed the Bonn Convention, which is properly called The Convention on the Conservation of Migratory Species of Wild Animals. Unfortunately (for butterflies) migratory birds were the lucky migratory animals to have regulations drafted in their favour. The few migratory butterflies of Britain (and Western Europe) were not included - although the definition of a migrant as 'any species or lower taxon of wild animal'..... which 'cyclically and predictably crosses one or more national jurisdictional boundaries' would seem to fit butterflies so well (cf. Lyster, 1985). The Bonn Convention has not been very helpful so far, for any conservation of butterflies in Britain. Rare species such as the Queen of Spain Fritillary, or the Short-tailed Blue, will have to wait for their moment of glory.

Then came another international agreement, CITES, which carries lots of importance regarding trade of rare species, but has very little to conserve British native butterflies. The UK signed the CITES agreement (the Convention on international trade in endangered species of wild fauna and fauna) in 1973, it came into force in the UK in 1976 and was further enforced with the EC Council Regulation of 1982 (EC Council Regulation, 1982). Unfortunately no British butterflies were put on any CITES list.

The import and export of named CITES species is normally restricted, and licences are issued by the Department of the Environment, where English Nature, Scottish National Heritage and the Countryside Commission for Wales act as advisors. It should also be said here that the international trade in tropical butterflies for the replenishment of 'butterfly houses' from outside Britain has to be done strictly in accordance to the CITES regulations, and that many *Papilio* swallowtails and birdwing (*Ornithoptera*) species which are attractive show-piece butterflies for butterfly houses are completely banned from importation. If the species concerned is captive bred or ranched in the country of origin, it stands a good chance of being licenced if the import will not be detrimental to the survival of the species.

The generally non-native butterflies kept in butterfly houses in Britain are beyond the scope of this book, but since Britain leads the world in butterfly enthusiasm and in the number of butterfly houses, which are part of the all-round pleasures of studying the conservation of butterflies, one is bound to take these import/export regulations seriously.

Most butterfly houses cater for the showy tropical butterflies; few cater for native British butterflies. Those that do include the Stratford Butterfly House (partially) and the Drum Manor Butterfly Garden set in an Irish country garden in Northern Ireland.

RAMSAR SITES

In 1979 the UK ratified the Ramsar Convention, properly known as the Convention on Wetlands of International Importance Especially as Waterfowl Habitat in 1973. Forty-eight Ramsar sites had been designated, with a dozen more under consideration, by the end of March 1994 (EN, 1994b). Designation of any Ramsar site will be excellent legislation for butterflies, since there are usually plenty of butterflies still associated with coastal areas, even some that are more and more restricted to coastal areas. As English Nature put it 'the UK government committed itself to promoting the conservation of particular sites and the wise use of wetlands within its territory' (EN, No.7, May 1993, p.4).

Most of the Ramsar wetlands are by their very nature coastal habitats, so that this designation will be inadvertently useful to two sorts of butterflies - not that they were considered for the designation; birds stole the Ramsar limelight in respect of wetlands, such as the Ouse Washes. First, coastal cliff-tops and headlands are often exceedingly important for a good range of browns and blues, as well as being receiving points / refuelling points for immigrant butterflies, but the butterfly with the most remarkable coastal distribution is the Grayling which may well stand to gain from some of the designated coastal sites. Where headlands are not intensively managed to the cliff edge and a corridor of suitable vegetation left, this is the best scenario for the perpetuation of coastal cliff butterflies.

The second sort of butterfly inadvertently protected through Ramsar are those that like wet habitats (often not too far from the coast). These include the Large Copper and Swallowtail whose populations are actively encouraged by people.

One must not forget that the international status of the Ramsar sites is mostly for bird conservation, and to this end, the habitat is regularly managed to maintain the preferred range of birds the public demands. The way that the land is managed for birds is often completely at odds with the management of the same land for butterflies. However, sometimes bird management is to the advantage of butterflies, especially on heathland habitats which are left alone. In recent years there has been much more active management for invertebrates, especially butterflies, on RSPB sites.

HABITATS AND SPECIES OF INTERNATIONAL STATUS

The UK Government encourages the voluntary sector to conserve habitats and wildlife and it will step in to save habitats or species of international importance. Two questions have therefore to be asked: are there any butterfly habitats in Britain of international importance, and are there are any British butterflies of international importance, both of which would thus have a good chance of being conserved by the UK government if the voluntary sector could not do it. It is a common tenet by government that if the voluntary sector is seen to be doing a good job in conserving nature, then there is absolutely no need to make available more funds.

Butterflies and the Law

It will be interesting to see in relation to the Habitats Directive how nature reserves for butterflies will be weighed in respect to butterfly diversity. Wicken might be designated for its habitat, but it supports only one key butterfly species, whilst a number of nature reserve on the Downs or Chilterns, which have 35 butterfly species, might never reach the statute books. And if they did, would they still be conserved? The search will obviously be on for UK butterfly sites that are of international importance, such as 'Wicken Fen Wetland Reserve' to give it its full title, which is described by the National Trust as 'the most important wetland reserve in Western Europe' (NT, 1986).

It is a sad reflection on the processes of nature conservation in Britain that the original international status of Dungeness SSSI (Kent), as a unique and historic place for migrant butterflies, has been progressively degraded and in parts completely eliminated. It was originally designated as international importance for its botanical, entomological and ornithological interest. The remarkable transformation (habitat change) of a significant part of the huge SSSI site, from an extraordinary expanse of natural gravel deposits into a freshwater wetland is unprecidented. Before conservation was as popular as it is today, planning consent was given for aggregate extraction on the site, with the approval, at the time, of the relevant conservation bodies. Now the emphasis has changed from its multi-interest appeal directly in favour of waterfowl, so much so that a remarkable range of unusual birds has started to nest there, giving it due cause for re-notification of SSSI status again. Exploitation of the gravel reserves on this SSSI Grade 1 site has occurred during the tenureship of the RSPB. This is not the only coastal 'bird landscape' created by the RSPB, another is on the North Norfolk coast at Titchwell in another conserved area, where lagoons have replaced coastal marshes which are useful refuelling stations for migratory butterflies.

EUROPEAN LAWS

There are no EC Directives specifically on the conservation of butterflies, as there are for birds such as the 1979 Conservation of Wild Birds. Of the 358 Environmental Directives, Decisions and Regulations issued from the EU up to November 1992, only the 'Habitats Directive' actually names butterflies in its statutes (DOE, 1992 Chronological List of EEC Environmental Directives, Decisions and Regulations). Although these Directives regulate the way the British countryside is managed to a greater or lesser extent, there are two groups of Directives from which butterflies profit, and have been profiting. These are Directives involving 'set-aside' and those to do with 'Environmentally Sensitive Areas' and 'Countryside protection in agriculturally less favoured areas'.

Butterflies and the Law

Land set-aside by farmers to reduce production in the EU has favoured butterflies since they have been able to colonize these formerly heavily worked soils, and exploit the increasingly wider range of grasses and other wild plants occurring in the fields. In set-aside land subject to minimal management such as 'topping' (cutting the weeds off at a certain height), butterflies have generally prospered. There were some original problems with the time of cutting in July, it being too early for ground-nesting birds, as well as for insects, but now cutting is recommended in August. Some bird sites designated under Special Protection Areas (SPAs) of which there were 28 designated in late 1992 (19 of which are both Ramsar and SPA sites) offer some other areas of habitat suitable for birds as well as for butterflies.

HABITATS DIRECTIVE

The Habitats Directive - as it is popularly referred to - is the Council Directive of 92/43/EEC of 21 May 1992 on the conservation of natural habitats and of wild fauna and flora. It was introduced to give those in the EC the opportunity to give legal force to the Bern Convention on the Conservation of European Wildlife and Natural Habitats which was signed in 1979 and came in force in the UK in 1982. The Habitats Directive was ratified by the UK government in 1993. Although the UK government signed and pledged agreement to a range of conservation activities, it decided in 1994 to go for derogation which means it prefers to apply its rules to the interpretation of the Directive.

The initiative for whole-habitat conservation, as embodied in the Habitats Directive, is a good one, and the protection and conservation of a wide range of habitats is an excellent ideal since this will help to conserve a wide range of species, including butterflies, specific to each habitat. The great problem with it at present is that it is highly biased towards continental habitats with few specific to Britain.

As its name implies the Habitats Directive is mostly concerned with habitats but it also has a check list of species to be conserved as well. As for species listed the Directive used a very similar list of butterflies compiled by the Berne Convention. Only 14 species of lepidoptera are listed for the whole of Europe and these include three moths and 11 butterflies. The lepidoptera mentioned in the Directive relate only to a minuscule percentage of the European lepidopteran fauna as well as of the British fauna (see Table 10.1). Three species on the UK list are specified, the Large Copper, Large Blue and Marsh Fritillary (Table 10.1) The reasons for the inclusion of the Large Copper and Large Blue are that they 'are rare and threatened with extinction' and, being 'indigenous to the Community, for which other Community legislation prohibits the taking or

Butterflies and the Law

trade'. The Marsh fritillary, the only UK butterfly listed which has not gone extinct, has experienced a 62% decline in Britain, and its colonies are being lost at 11% per decade (BC Marsh Fritillary Fighting Fund, 1994).

Table 10.1 Butterflies listed in the Habitats Directive.
(using original spelling of Latin names in document)

Common name	Latin name
ANNEX II a)	
Animal and Plant species of community interest whose conservation requires the designation of special areas of conservation.	
Marsh Fritillary	*Euphydryas aurinia*
Large Copper	*Lycaena dispar*
ANNEX IV a)	
Animal and Plant species of community interest in need of strict protection.	
Large Copper	*Lycaena dispar*
Large Blue	*Maculinea arion*

One cannot help feeling that some other deserving species must have been overlooked. Extant species are infinitely preferable to extinct species. Is it perhaps a reflection on lepidopterists as field workers and observers, and academics and government agencies responsible for advising the government, that the case has never been put strongly for the protection of the fragile populations of British butterflies.

To warrant inclusion onto Annex II of the Habitats Directive means that Special Areas of Conservation should be set aside for their protection. Species within Annex II that are attributed with an asterisk are given priority status; sadly, neither the Marsh Fritillary nor the Large Copper has been afforded this extra protection. Perhaps this pecking order within Annex II will eventually be used against them in a court of law.

Getting the whole package of habitat protection throughout Europe off the ground will not be easy, and there are already signs of a certain degree of reticence and lack of commitment often associated with Community regulations. The UKs' Consultation Document was finally published on 4 October 1993. The UK committed itself to getting all the legislation and administration necessary to fulfil its obligations under this Directive to effect habitat designation by June 1994.

By June 1995 the UK must submit to the Commission a list of sites (not all sites) containing habitats or species listed in Annexes I and II of the Directive. These will be sites eligible for selection as Sites of Community Importance

Butterflies and the Law

(SCI), which, in the UK, will be called Special Areas for Conservation (SACs). By June 1998 the Commission will publish the list of SCIs, and by June 2000 the first of the 6-yearly implementation reports should be published.

At the time of writing (April 1995) specialists both within and outside the statutory agencies have submitted confidential lists of proposed SAC sites for habitats and species. The Joint Nature Conservation Committee (JNCC) -an organisation which speaks for several conservation organisation - will have examined all these proposed sites and they will put forward a list of about 200 sites for designation in England. Other similar sites will be designated in Wales, Scotland and Northern Ireland to make a total figure of about 400.

The total list of SACs in Britain will not be available until after publication of this book. It is anticipated that by the end of 1994 it will be possible to go public with the draft 'confidential' list, only after all the owners and managers have been contacted and canvassed. Some of the sites are in multiple ownership or are run as Trust-funds, or run by institutions or commercial organizations, as well as having tenants and other right-holders. The task of informing everyone is colossal, and the chance of them all agreeing to publication is remote.

HABITATS

The range of habitats to be protected throughout the width and breadth of the EU is quite generous, and one hopes that every worthy example of each of the many and various habitats outlined under each category is offered protection under this Directive. If any of these sites is protected and eventually afforded SCI status, then ALL butterfly species occurring in these designated habitats will theoretically enjoy some increased degree of conservation and protection. Examples of the sorts of habitats that may receive due consideration in the UK are given in Table 10.2. One would be forgiven for thinking that the 'Habitats Directive' is drafted with alpine and Mediterranean habitats mostly in mind; under some categories it is impossible to find appropriate descriptions to encompass common British habitats.

Table 10.2 Habitat types specified in Habitats Directive, typical examples of which may be found in the UK.

Coastal habitats: glasswort meadows (15.11) eg. North Norfolk coast
Coastal habitats: machair (1.A) eg. West Scotland
Coastal habitats: humid dune slacks (16-31-35) eg. Studland, Dorset
Freshwater habitats: turloughs (Ireland) (none) eg. County Clare, Eire
Temperate heath and scrub: North Atlantic wet heaths (31.12) eg. Lizard heathland

Butterflies and the Law

Sclerophyllous scrub: Juniper on heaths & grasslands (31.88) eg. N. Downs, Chilterns
Natural & Semi-natural dry grasslands: calcareous grasslands (34.31-34)
eg. N. & S. Downs
Mesophilic grasslands: lowland hay meadows (38.2) eg. Cricklade SSSI, Wiltshire
Mesophilic grasslands: mountain hay meadows 'British types with *Geranium sylvaticum*
(38.3) eg. British uplands
Raised bogs, mires & fens: alkaline fens (54.2) eg. Wicken Fen
Raised bogs, mires & fens: calcareous fens with Cladium sedge (53.3) eg. Wicken, Woodwalton
Raised bogs, mires & fens: eutrophic tall herb communities (37.7,37.8) eg. Wicken, Woodwalton
Forests: old oak woods with holly, hard fern..(41.53) eg. many sites in south-east England

Note: The habitat type described is shortened from the official description which can be consulted in the actual 'Habitats Directive' document, and the habitat codes shown in column 3 are for those who wish to verify the actual type of habitat considered under this Directive.

Any listed butterfly species that happens to fly in a special type of habitat so designated by the Directive will be automatically conserved. The wording in the Consultation Document states: 'Any site on the national list hosting habitat or species considered to be a priority by the Community (identified by an asterisk in Annex I or II) will automatically be selected as a Site of Community Importance'. These priority habitats are those that are in danger of disappearing and have higher priority than other commoner types of habitat. Thus any Marsh Fritillary that flies in any special site such as a turlough will have to be protected. To explain more fully, one of the special habitats cited in the Habitats Directive is 'Turloughs (Ireland)'. Turloughs are mostly typical of the Republic of Ireland, but there is a single turlough in Wales (in Carmel Woods, Dyfed). However, the latter site does not have any Marsh fritillary so this designation is not at issue in the UK. (1) The conservation of Marsh Fritillaries would be more of an issue if, for instance the Marsh Fritillary populations in the Burren were to be found on any of the turloughs of County Clare, Eire.

The Large Copper is also a listed species on the Habitats Directive, but there are no priority habitats listed on the Directive. Under the category 'Freshwater Habitats' there are two sections 'Standing Water' habitats and 'Running Water' habitats. The Large Copper exists at Woodwalton Fen, which is a running water type of habitat (but strictly controlled by people) and it may eventually fly in the Broads (running water). There appears to be no appropriate habitat described for the Large Copper in the Directive, either under running water (where there are three alpine habitats and five Mediterranean / mountainous types of habitats described) and there are no appropriate habitat labels to include it under 'standing water'. The dilemma is that the Large Copper, as a designated species, has to be protected, but its current and projected habitats do not appear to be accommodated by the Directive.

CONSERVATION STATUS

The conservation status of butterflies, as represented by the system of Red Data Book (RDB) categories, is very important in assessing their well-being. RDB status does not have any legal basis, but it is used as a foundation for establishing British wildlife law. Some discussion of RDB status and butterflies therefore follows.

Butterflies requiring protection need a tag - a sort of benchmark - by which their current health status can be assessed. Once pigeon-holed into a certain category conservationists are then able to apply (hopefully) a known set of remedial actions to look after that species.

Fortunately the International Union for the Conservation of Nature (IUCN) drafted a set of status criteria which are usefully applied worldwide and help in interpretation and comparisons; these can be seen in the glossary and are discussed here. The slippery slide to the extinction of a butterfly is well charted in appropriate terminology. A living species is known as extant, yet it begins its path to extinction by going through the following categories of status: rare, vulnerable, endangered and finally extinction in that order. The word threatened is not a word to use, in Europe at least, since it is not afforded any precise definition. The term is used, however, in the USA, where it does have a very strict definition, but that is beyond the scope of this book.

These status categories are often used in Red Data Books which cover most groups of plants and animals, usually on an international level. There are no IUCN Red Data Books specific to butterflies of Britain, but there are two RDB books which mention a few butterflies species in Britain (Collins & Morris, 1985; Shirt, 1987). The concept of Red Data Books is an excellent one especially for planners and conservators since they have a ready reference to species which are in need of preservation or conservation. A butterfly that is included in a Red Data Book is given some degree of importance, of a quantitative assessment of its fortunes, the data for which are sometimes lacking in other species. Although many of the Red Data Books have an international focus, it is pleasing to report that there is a growing literature of books listing RDB species on a county by county basis (e.g. Mahon & Pearman, 1993; see Table 10.3).

Table 10.3 County status of butterflies in Dorset

Adonis blue S
Brown hairstreak S
Duke of Burgundy S
High Brown Fritillary R
Lulworth Skipper S
Marsh Fritillary S
Pearl-bordered Fritillary D
Purple Emperor S
Silver-spotted Skipper R
Silver-studded Blue S
White-letter hairstreak D
Wood White S

Key R = National Red Data Book Species (occurring in fewer than 16 x 10 km squares nationwide). This 'R' category is a combination of the three national red data data list classifications, RDB1, RDB2 and RDB3. S = Nationally Scarce Species (occurring in 16 to 100 x 10 km squares nationwide). D = Dorset Scarce Species (not in category 1 or 2 and occurring in three or fewer sites in the county)
Source: Mahon & Pearman, 1993.

Unfortunately a species conferred with RDB status is not supported by any legislation to conserve it. RDB status is used, and has been used, as a measure to draw up lists of species that deserve to be listed in wildlife acts. There are different categories that IUCN RDB species qualify for, from Category 1 (breeding in internationally significant numbers) to 4 (localized non-breeder). Category 5 is a special category for those whose status gives cause for concern, such as those showing declining numbers, but for which there are inadequate data available. There are also candidate species lists, of those species that might one day gain RDB status.

Lest there is confusion between IUCN and British categories of RDB status, those of the British RDB status are the following: Category 1 Endangered is 'Taxa in danger of extinction and whose survival is unlikely if the causal factors continue operating'; Category 2 Vulnerable is 'Taxa believed likely to move into the Endangered category in the near future if the causal factors continue operating'; Category 3 Rare is 'Taxa with small populations that are not at present Endangered or Vulnerable, but are at risk'; Category 4 is 'Taxa formerly meeting the criteria of one of the above categories, but which are now considered relatively secure because effective conservation measures have been taken or the previous threat to their survival has been removed'; Category 5 is 'Taxa which are not known to occur naturally outside Britain' (Shirt, 1987).

Butterflies and the Law

Shirt's book gives a comprehensive account of these categories, and also notes that 21.4% of British butterflies are afforded RDB status.

The status of Britain's butterflies in the British RDB, as listed in Shirt (1987) is as follows: RDB Category 1 includes the Large Blue and the Large Tortoiseshell; RDB Category 2 includes the Heath Fritillary, High Brown Fritillary and Swallowtail; RDB Category 3 includes the Glanville Fritillary and Silver-spotted Skipper; RDB Category 4 includes the Chequered Skipper and Black Hairstreak (both now out of danger) and RDB Category 5 (endemic races) includes two, the Grayling, *Hipparchia semele thyone*, and the Silver-studded Blue, *Plebejus argus caernensis*. These last two species are well-known endemic races, confirmed as major subspecies. Other details about these species and how their populations can be assessed in the planning process can be found in NCC, 1989.

In respect of SSSI designation of a site, the NCC (1989) actually relied on RDB status as a major criterion for consideration, although they stressed that applying a scoring system to butterflies in Britain is fraught with difficulty. The problem arises from the disparity of butterfly species represented along the length of Britain - 40 species in some southern English counties and ten in the far north of Scotland. This disparity of distribution plays havoc with any assessment or scoring system which should be applied equally across the country. In the opinion of the NCC (1989) there are just two 'nationally rare species of butterfly', the Large Blue (RDB category 1) and the Large Copper which is in the RDB appendix. The 'rare' here is not used in the strict RDB sense. Both are listed as extinct in NCC (1989) p.281.

Mark Collins & Mike Morris' book deals with the threatened swallowtails of the world, an important family of butterflies, but gives only a passing comment on the British Swallowtail, with its 37 worldwide subspecies, one of which is the English *Papilio machaon britannicus* (Collins & Morris, 1985). This British subspecies is regarded as 'not threatened' throughout most of its range in England and in the world, although it is protected by law in various European countries. The status of *britannicus* is not thought to have changed since it was so designated by Collins & Morris (1985).

UK LAWS

The first British butterfly to receive total protection in law was the Large Blue which was listed in the Conservation of Wild Creatures and Wild Plants Act 1975. Since then it has been joined by precious few butterflies, although many butterfly populations generally have declined alarmingly.

More comprehensive laws protecting UK butterflies were laid down in the Wildlife and Countryside Act 1981. This was a troublesome Act when it appeared since it was drafted in a very ambiguous manner and was difficult to interpret because of the numerous cross-referencing (cf. Stubbs, 1991).

Just four butterfly species were given full protection in the 1981 Act. One of these had become extinct in Britain a few years earlier, the Large Blue, and the Chequered Skipper had suddenly disappeared permanently from its English haunts during the same period. The Heath Fritillary and the Swallowtail are both southern butterflies isolated in fragments of suitable habitat. Legislation conservation was therefore targeted at specific species with fragile populations, often living in habitats well managed by humans. There were no provisions in the 1981 Act for any common butterflies, or for habitats or nature reserves with a rich suite of species.

The Chequered Skipper was removed from the 1981 published list in 1986 after the NCC advised that it no longer needed full protection (Stubbs, 1991). It now joins a list of 22 scarce native butterfly species whose trade (i.e. buying and selling) is now controlled and monitored. Species on this list can be taken by individuals in the wild, but the ...'sale of any specimens of the 22 partially protected and three fully protected species is illegal unless: a. the specimens were bred in captivity or b. a sale licence issued by the Department is held.' (Stubbs, 1991). (The partially protected species are the Northern Brown Argus, Adonis, Chalkhill, Silver-studded and Small Blues, Large Copper, Purple Emperor, Duke of Burgundy, Glanville, High Brown, Marsh and Pearl-bordered Fritillary, Black, Brown and White-letter Hairstreaks, Large Heath, Mountain Ringlet, Chequered, Lulworth and Silver-spotted Skippers, Large Torotiseshell and Wood White). That there is a clampdown on commercial collecting is a little curious since most professional lepidopterists argue strenuously that collecting has had little effect, when, in fact the *bête noire* of butterfly decline is habitat loss. This is only partially addressed in the Habitats Directive.

Has legislative conservation of the four species protected under the 1981 Act worked? The first successful prosecution occured in 1994 when two men were found guilty for possession of wild Chequered Skippers offered for sale at an Entomological Fair at Granby Hall, Leicester in 1993 (RSPB, *Legal Eagle*, Winter 1994, No.5, p.4). One person was fined £490 costs for possessing 14 wild Chequered Skippers for sale, the other, £80 costs for offering seven wild

Butterflies and the Law

Chequered Skippers for sale. This case makes history in the world of legislative butterfly conservation. Up to now the law supporting the protection of four species which have full protection including the protection of the others in respect of sale, had not been fully exploited. It was found to be wanting. Perhaps now there may be some who would argue strongly for stiffer penalties, even related to the number of individual butterflies (or their life stages) having been taken from the wild. This would support the law as it stands now. Perhaps there will now be stricter administration of the 1981 Act, and vigilance by members of the public after the general awareness that this case has given.

The Large Blue conservation story continues to be a top secret affair, conducted off limits in various localities in the West Country, and as such, is a strategy which protects both the site and the species from possible depredations of collectors and enthusiasts. The Chequered Skipper seems to be holding its own in Scotland with healthy populations relatively untouched by collectors. Both the Heath Fritillary and the Swallowtail do not seem to be under any great threat, the Heath Fritillary living in shifting populations in various managed woodlands, and the Swallowtail having the advantage of flying in relatively inaccessible habitats.

More significant protection for butterflies (all butterflies, not just the listed ones) is found under the SSSI regulations also embodied in the Wildlife and Countryside Act 1981. Adequate provision for the protection of butterflies is embodied in the Act, a fact re-iterated in the Consultative document published in 1993 on the Habitats Directive.

HAS SSSI PROTECTION WORKED FOR BUTTERFLIES?

No not 100%, it is not a safeguard, and cannot be relied upon. The SSSI structure is meant to be a fail-safe method of conserving the best range of different habitats in the UK - the cream of the countryside with all the attendant flora and fauna to offer. Unfortunately it cannot be relied upon to safeguard this reserve collection of British wildlife, let alone butterflies, although the 80,940 ha of SSSI-land on NT properties, representing one third of their 560 properties, is likely to fare much better than SSSI-land not in NT hands (NT, 1992).

All is not as bad as it may seem, however. A letter sent to all the 47 RSNC Wildlife Trusts asking them for information on loss of butterflies from SSSIs resulted in comments from only two of the Trusts; one from Wales concerned threats to Marsh Fritillary populations, the other from Dorset three sites saved on nature conservation grounds: Lodmoor SSSI (Weymouth), Canford Heath SSSI (Poole) and Bourne Bottom SSSI (Poole). Amongst the butterflies saved on these sites were Marbled Whites, Graylings and Silver-studded Blues. Frequent comments from the Wildlife Trusts were that there were examples of lack of management resulting in the decline of butterfly colonies. Staffordshire and County Durham Trusts explained that lack of management, causing scrub encroachment, is a significant problem in conserving butterflies in species-rich grasslands.

There have been other butterflies saved on SSSIs. One such site is at Whixall Moss SSSI in North Shropshire. Jenny Joy and Mike Williams originally drew attention to the plight of the Large Heath on this stretch of wilderness in 1990 (Joy & Williams, 1990). The site is noteworthy since this is the southernmost edge of the Large Heath's English range and both the subspecies *davus*, and *cockaynei* occur here; the butterfly also breeds here on Hare's-tail Cotton Grass, *Eriophorum vaginatum* which is not always listed as the typical foodplant. The site was under threat from a four-fold increase in commercial peat digging and there was virtual outline planning permission existing. A campaign to save Whixall Moss was successful at bringing awareness of the site to the general public as well as to scientists and planners since there were 11 RDB species of invertebrate recorded there. The result was a revocation of the planning permission and the NCC declared the site a NNR giving it more protection. It is now run by English Nature and the Countryside Commission for Wales who are raising water levels. The future of Whixall Moss now looks secure. The local rights to hand-cut peat-digging continue in a small way, which is beneficial to the butterflies, since it allows opening up of the peat and re-colonization by flora and fauna. There is another Large Heath locality in the region, at Wem Moss, a Shropshire Wildlife Trust reserve, which is isolated in an agricultural environment. Its future is less secure.

The system of SSSI protection is well-known to be flawed, since it is open to unregulated abuse and degradation (cf. Wilkinson, 1990). Wildlife Link, an organization comprising most of the conservation bodies in the UK, commissioned Dr T. A. Rowell to investigate the state of SSSIs in the country, and his resultant findings are disturbing. In the year prior to the 1981 Act, 13% of SSSIs were estimated to have been damaged comprising 8700 ha affected and 2400 ha beyond recovery. Things have got a little better since then, but 'around 5% of SSSIs are recorded as being damaged in any one year during routine loss and damage recording'. Rowell says this is an underestimate. He visited one-third of SSSIs in England in 1990 under the Site Integrity Monitoring Scheme and found that 40% of the sites showed deterioration or damage, and 21% were under threat. His report makes sobering reading and bodes badly for butterflies, as well as the rest of the cream of British wildlife. Random damage of rich wildlife habitats continues.

Butterflies and the Law

Where any SSSI is damaged or lost butterflies are likely to be affected since they are found in most of the habitat types represented by the various SSSIs in Britain. The sad thing about butterfly loss from SSSIs (and of course other parts of the countryside) is the piecemeal way in which it happens. Humans are often to blame either directly or indirectly. Key butterfly species might not have been eliminated from known SSSIs, but the loss of sensitive species from SSSI habitats has happened across the country by a variety of means.

It would be convenient if key butterfly localities occurred in SSSIs, then there would be a marginal chance of them being safeguarded. Carmarthenshire, for example, is fortunate in having a lot of its grassland sites within the series of SSSIs, particularly in the Dinefwr Borough (Morgan, 1989). But an SSSI is no guarantee of protection. Carmarthenshire has not had the high level of meadow loss experienced through most of Britain, that is, between 95 and 97% of meadows lost (Feltwell, 1992) but it is losing some of its finer butterfly strongholds, particularly those of the Marsh Fritillary because they occur in the same area as the open cast coal mining. Carmarthenshire is particularly important for meadows and Marsh fritillaries, since the pasture colonies of Dyfed are important in a national context (Morgan, 1990). Just where the shales occur close to the surface is ideal both for butterflies as a breeding site (because of the wet conditions produced) and open cast mining. So there is an increasing conflict between coal extraction and butterfly conservation. Restoration of worked-out mines, so far, has not concentrated on conserving butterflies, rather on returning the land to 'improved' agriculture or forestry (Morgan, 1989). Where there are still many unimproved meadows and butterflies, such as in the upper Gwendraeth Fawr and Amman Valleys with the connecting Cross Hands and Ammanford area, there are rich seams of coal ideal for open-cast mining. The conservation prospects for the Marsh Fritillary do not look particularly good, whether it occurs on an SSSI or not, and despite it being now of national and international importance.

The general attrition of SSSIs, by the 'improvement' of the land, and by the deliberate destruction of SSSIs before notification, has caused the extinction of butterflies. This has occurred regularly and widely, and has not often been reported because it is piecemeal. In parts of the West Country populations of Small Pearl-bordered Fritillaries and Silver-studded Blues have disappeared from some SSSIs (Warren, pers. comm. 1993) There are also two other worrying aspects of butterflies disappearing from SSSIs. First, the lack of management of SSSIs has in some cases caused the loss of populations, and, second, some important sites, such as Bernwood Forest SSSI is still losing species although all concerned parties (Forestry Commission and local trusts) have been doing their best to conserve this historic butterfly haunt. The future of butterfly populations on SSSIs does not look good. It is the general attrition of the populations, their fragmentation and isolation, that is at the root of the ecological and conservation problem.

It is always a sad fact to report that loss of habitats through neglect caused the demise of certain butterflies, but this is a frequent scenario. Habitat loss was the case with the Large Blue, and it is may be the case with the Marsh Fritillary. A recent report on the traditional 'rhos' pastures of Carmarthenshire showed that about 50% of surveyed sites exhibited habitat degradation, especially in Cefn Blaenau and Tir Philip (Morgan, 1992). Marsh Fritillaries depend on a delicate balance of grazing and cutting to maintain the older pastures, and if this is not done, then rank grasses invade only to be replaced by scrub; 'Put bluntly, sites have to be grazed annually at a regular stocking density which equates to approximately 0.7 livestock units (or one pony per hectare)' (Fowles, 1991). When this is done successfully, then big populations of Marsh Fritillaries may result - for a place like Rhos Llawr-cwrt (Ceredigon) in 1991, there were estimated to be about 20,000 Marsh Fritillaries emerged, making this one of the largest populations of this insect in Britain.

NATIONAL NATURE RESERVES (NNRs)

National Nature Reserves are often substantial tracts of land which have been designated by the government agencies (English Nature, Scottish National Heritage, Countryside Commission for Wales) to protect particularly fine examples of habitat. Although large, and often containing a suite of different habitats, they are usually isolated in the countryside, and certainly isolated from each other. They therefore lack something of a continuum which can be found in some coastal SSSIs, such as along the Cornish coast. In the context of corridor conservation and coastal butterflies this is highly relevant. The purpose of this section is to see how NNRs are regulated and how butterflies fare in them. The system of NNRs in England represents some of the finest habitat types in Britain, sometimes in Europe, and affords them statutory protection. But are butterflies safe in NNRs and are they being conserved well?

Theoretically butterflies in NNRs should be protected and safe, since NNRs are owned, leased, held or managed by an approved body such as EN, SNH, County Trusts. By the end of March 1993 there were 140 NNRs in England covering 57,000 ha (EN, 1993c). Forty-five NNRs are owned entirely by English Nature, so these MUST, one assumes, be safe in their hands. For further details on how NNRs have been set up, readers should consult Morris (1967); suffice it to say that statutory nature reserves include LNRs and NNRs as well as NNRs. Many NNRs are leased and about 40 are leased with an Nature Reserve Agreement.

Butterflies and the Law

No figures for loss or damage to NNRs are quoted by English Nature (unlike data for SSSIs which are available elsewhere). However, as all NNRs are now SSSIs (Sheppard, pers. comm. 1994) there is likely to be some damage to them since they would be represented (however small a percentage) within the annual 5% loss and damage to SSSIs, supposing this to be continuing each year. In the latest annual report from English Nature it is noted that damaging incidents affected 117 SSSI to 31 March 1994 (English Nature, 1994).

On a rather important historical note Alan Stubbs (1979) who was always a staunch promoter of conservation matters whilst he worked for NC and NCC, was also their strongest critic. He recalled how the NCC 'got the wrong piece of wood at Ham Street (Kent) as an NNR in the 1950's' .... and 'got the wrong piece of wood at Blean (Kent)' for various reasons including lack of money and the slump in coppicing'. It was here that the Heath Fritillary died out on the NNR.

But there have been other surveys of butterflies on NNRs. Warren (1992c) has given details about these: 11 species have become extinct from Monks Wood NNR since it was established in 1953 when it had 35 species; and 13 species lost from Castor Hanglands since it was established in 1954.

On the same subject Stubbs stated ' Foremost, we have failed to halt the decline of the Large Blue' .....'the Large Copper.... was not firmly established at Woodwalton Fen ' and.....requires a zoo culture to keep it going'.....'The spectacular crash in butterfly diversity at Castor Hanglands and Monks Wood is widely known and the Chequered Skipper has virtually (?completely) died out in England'. Despite these losses from NNRs and other reserves, the track record of conserving butterflies on NNRs has been relatively successful.

Scientists at Monks Wood monitor the conservation of various insect groups in Britain, and often use the network of NNRs as an indicator of the health of the nation's wildlife. The NNRs are also used as a basis for the long-term monitoring of butterflies; the Butterfly Monitoring Scheme was set up by Ernie Pollard when he worked at Monks Wood. The fact that NNRs are relatively safe and stable from environmental impact means that long-term recording from NNRs is more meaningful than having to rely on other parts of the wider countryside.

A major review of the butterflies on British Statutory Nature Reserves was published in 1967 by Mike Morris (Morris, 1967a,.b), but there appears to be nothing since. Findings indicated that butterflies were almost certainly found in most of the NNRs since many were so large, but data from NNRs were generally lacking: 'Few statutory nature reserves have been established primarily for entomological reasons'.. 'In a reasonable number of reserves, the entomological interest is important'. Good data are present from just a few NNRs (e.g. Rhum NNR, Woodwalton Fen NNR).

That a third of the NNRs by 1967 had not a single butterfly recorded present is, on the face of it, an alarming figure. It highlights, once again, the paucity of records on butterflies. Our ignorance of their status in the countryside is often reflected in their demise. Another alarming conclusion from the Morris report was that three of the more interesting species, the Large Blue, Lulworth Skipper and the Glanville Skipper were not found in any NNR in Britain. The Large Blue was to become extinct from the English countryside in under two years from this survey, and it became extinct without a statutory nature reserve ever having been created for it - not that at least 50 years of intensive study on a flagship species was not enough to understand the pattern; and to heed the warnings. Morris ended his review of butterflies on NNRs by saying 'It is very much to be hoped that in another decade, it will be possible to write much more authoritatively on the representation of butterflies on our statutory nature reserves.' The situation now is that the Large Blue will be present on about ten sites in the West Country, none of which is designated as an NNR (Sheppard, pers. comm. 1994). That the Large Blue is conserved on non-NNR land, some of it owned by the NT, is also good, however since the NT have a good track record of protecting wildlife in their charge. Neither the Glanville Skipper, in its stronghold on the Isle of Wight, nor the Lulworth Skipper, confined to the Dorset coast, is protected within an NNR, though they exist in coastal areas which do have linear SSSI conservation.

The question is quite rightly raised as to whether rare butterflies need to be conserved within statutory nature reserves, or even non-statutory nature reserves. Do butterflies really need to be conserved in NNRs? Are not the other thousands of nature reserves sufficient for their long-term protection?. The NNRs are, after all, a set of habitats, not at all selected for butterflies, let alone insects, which is used as some form of basis for conserving and monitoring Britain's flora and fauna.

NNRs are generally safe havens for butterflies and one would expect lots of research to be going on in them. In 1965 Duffey and Morris noted that only 47 projects out of 550 for that year involved invertebrates in NNRs. The latest English Nature Research Programme 1992/3 (EN, 1993b) mentioned only two NNRs by name Old Winchester Hill NNR and Chippenham Fen NNR, of which only the former had three projects involving surveying invertebrates. There were slightly more projects on SSSIs (than NNRs) mentioned for 1992/3 yet none of these involved butterflies.

Current data indicate that butterflies are being actively researched on about 70 of England's NNRs, and that most of these concern counts of butterflies along transects (Table 10.4). Most of these will involve the BMS which records details of butterfly abundance. It is good to see a nice range of butterfly species mentioned: Marsh Fritillary, Brown Argus, Large Heath, High Brown Fritillary,

Butterflies and the Law

Orange Tip, Large Copper and Heath Fritillary.

The 1960s and later data on butterfly research in NNRs are not directly comparable, because it is not always possible to sort out butterflies from invertebrates, and for geographical regions. The earlier high figure of projects may be because Welsh and Scottish sites were included, and becasue the then Nature Conservancy had direct responsibility for NNRs and research. Now the work on NNRs is done by country agencies, the ITE and other research bodies, so figures for project work on NNRs from the EN do not include these field data.

Table 10.4
Research and monitoring on National Nature Reserves in England.

Abstracted from the NNR/MNR Report, 1992/3 collated by Eddie Idle, and reproduced here by kind permision of English Nature (English Nature, 1994). The sites are listed according to NCC Regions.

National Nature Reserve, County, Research being undertaken

Ainsdale Sand Dunes NNR, Cumbria. Lepidoptera research
Finglandrigg Woods NNR, Cumbria. Marsh Fritillary monitoring
Gait Barrow NNR, Lancashire. Butterfly research; project to determine optimum
    conditions for High Brown Fritillary
Glasson Moss NNR, Cumbria. Monitoring invertebrates in response to bog restoration
Moor House NNR, Cumbria. University research
Roundsea Woods and Mosses NNR, Cumbria. Butterfly monitoring
Thorne Moors NNR, Yorkshire / Humberside. Invertebrate monitoring
Castle Eden Dene NNR, Durham. Butterfly transect
Ingleborough NNR, Lancashire. Preparation of Lepidoptera data base
Lindisfarne NNR, Northumberland. Butterfly transect
Thrislington NNR, Durham. Durham Brown Argus monitoring butterfly transect (Sam Ellis of
    Houghall Agricultural College)
Upper Teesdale NNR, Durham. Butterfly transect
Aqualate Mere NNR, Staffordshire. Invertebrate survey
Chaddesley Woods NNR, Herefordshire and Worcestershire. Butterfly transect
Derbyshire Dales NNR, Derbyshire. Butterfly transect
Fenns, Whixall & Bettisfield Mosses NNR. Shropshire   Large Heath surveys.
    Field Studies Council - Insect monitoring baseline.
    Liverpool Museum - insect survey; Butterfly transect
Moccas Park NNR, Herefordshire. Insect Survey
Rostherne Mere NNR, Cheshire. Butterfly transect
Wren's Nest NNR, West Midlands. Butterfly transect
Wynbury Moss NNR, Butterfly transect

Butterflies and the Law

Wyre Forest NNR, Herefordshire & Worcs. Woodland fritillary monitoring, BC transect.
Barnack Hill & Holes NNR, Cambridgeshire.   Butterfly transect
Barton Hill NNR, Bedfordshire. Butterfly transect
Buckingham Thick Copse NNR, Buckinghamshire. Tony Warne - insect response to ride management (private research by JNCC staff member) Butterfly transect
Bure Marshes NNR, Norfolk. Butterfly transect
Castor Hanglands NNR, Cambridgeshire. Butterfly transect
Chippenham Fenn NNR, Cambridgeshire. Butterfly transect
Collyweston Great Wood NNR, Northamptonshire, Butterfly transect
Colne Estuary NNR, Essex. Butterfly transect
Dersingham Bog NNR, Norfolk. Butterfly transect
Gibralter Point NNR, Lincolnshire. Butterfly transect
Hickling Broad NNR, Norfolk. Butterfly transect; Swallowtail monitoring
Holkham NNR, Norfolk. Butterfly transect
Holme Fen NNR, Cambridgeshire.   Butterfly transect
Monks Wood NNR, Cambridgeshire. Butterfly transect, Rare butterfly monitoring
ITE - Orange tip butterfly and lady's smock project
Saltfleetby-Theddeldthorpe NNR, Lincolnshire. Butterfly transect
Scolt Head NNR, Norfolk. Lepidoptera recording
Swanton Novers NNR,   Norfolk. Butterfly transect
Upwood Meadows NNR , Cambridgeshire. Butterfly transect
Walberswick NNR, Suffolk. Butterfly transect
Weeting Heath NNR,   Norfolk. Butterfly transect
Winerton Dunes NNR,   Norfolk.   Butterfly transect
Woodwalton Fen NNR, Cambs. Butterfly transect. Keele Unvi. Large Copper research
Avon Gorge NNR, Avon. Butterfly transect
Barrington Hill NNR,   Somerset.   Butterfly/moth count
Bovey Valley Woodlands NNR, Devon. Butterfly transect
Ebbor Gorge NNR, Somerset. Butterfly transect
Golitha Falls NNR, Cornwall. Butterfly transect
Hambledon Hill NNR, Hampshire/Dorset. Butterfly transect
Hogg Cliff NNR, Dorset. M.S. Warren grassland and scrub for butterflies
Rodney Stoke NNR, Somerset. Butterfly transect
Shapwick Heath NNR,   Somerset.   Lepidoptera counts. Various insect counts
Stoborough Heath NNR, Dorset/ ITE research on heathland restoration for invertebrates
Studland Heath NNR, Dorset. Butterfly transect
Yarner Wood NNR, Somerset. Butterfly transect
Beacon Hill NNR, Hampshire. M.S. Warren - butterfly research
Chimney Meadows NNR, Oxford Museum - invertebrate survey
Cothill NNR , Oxfordshire. Invertebrate recording
Martin Down NNR, Hampshire. ITE - effect of chemicals on chalk grassland invertebrates

149

Butterflies and the Law

Old Winchester Hill NNR, Hampshire. ITE - collation of earlier invertebrate recording and research for provision of management advice
Wychwood NNR, Oxfordshire. Invertebrate recording
Blean Woods NNR, Kent. Status of Heath fritillary butterfly transect
Castle Hill NNR, Kent. Butterfly transect
Ham Street Woods NNR, Kent. Butterfly transect
High Halstow NNR, Kent. Butterfly transect
Lullington Heath NNR, East Sussex. Butterfly transect
Mount Caburn-Lewes Downs NNR, East Sussex. Butterfly transect
Stodmarsh NNR, Kent. Butterfly transect
Swale NNR, Kent. Butterfly transect

Probing a little more beneath the surface of EN and the work carried out by them and for them, it appears that there is slightly more work being carried out on butterflies than at first envisaged. Martin Drake at EN has carefully gone through hitherto unpublished data returns from NNRs, especially a report by Peter Croucher on invertebrate population monitoring in Britain and Ireland (which excluded data from the BMS), and from returns made from site wardens of NNRs. The NCC used to collate information from all the British NNRs, but since the NCC split, this has been discontinued (a sad day for comparative records). According to the latest information from EN at Peterborough, there were butterfly transects being carried out on 53 NNRs in England and another 28 transects which were not included in the ITE scheme. (2)

The outlook for the conservation of habitats as assessed by Alan Stubbs in 1981 was pretty grim, and that was before the NCC split, which made nightmares of administration and assessment of comparative data worse. He called for a review of management of policies of the then 167 NNRs, and for an assessment for the 800 sites described in the masterly *Nature Conservation Review* (Ratcliffe, 1977). Most conservation sites are in private hands, and the annual loss of SSSIs is running at a conservative 5% per year, 'compounded over a few years, the outlook is catastrophic.' That was not all; Stubbs also drew attention to the NNRs being mostly on lease from private owners and that in the current economic climate (that was in 1981 - and it has worsened since then) many of these leases are coming up for renewal and they may be resisted. On a more optimistic note, it is pleasing to report that most NNRs now have management plans either finished or under way which will help to conserve Britain's butterflies. This is now a very significant step forward in the conservation of Britain's butterflies.

Butterflies and the Law

Notes

1. Turloughs are transiet lakes on limestone which come and go very rapidly depending on the vagaries of the water table. Typically a feature of the Burren in County Clare, Eire, they are supported by a network of underground caves. Because the land is highly fertile after the water has disappeared, the grass grows strong and is grazed by cattle; wet patches remain which are suitable as reserves for butterflies, such as the Marsh Fritillary. The site in South Wales is Pant-y-llyn lake near Carmel; a recent important study of the invertebrates there, however, did not touch on lepidoptera (Rundle, S.D. (1993) in Bulletin of the British Ecological Society, Vol. XXIV (4) 215-221.)

2. This information was gleaned from Idle's and Croucher's report, and I am grateful to Martin Drake at EN for providing it.

CHAPTER ELEVEN
VOLUNTARY CODES AND PRACTICES

Even in such wild places as the Lancashire mosses, when I have been collecting *Coenonympha tiphon,* the swallows would take two out of three that rose on the wind when the sun shone.   S.G. Castle Russell, 1925

INTRODUCTION

Legislative conservation can go only so far. And as far as it has been so far is to make it illegal to collect any of three butterflies on the UK list, or to disrupt their habitats. Up to 1994 there had been no person fined or imprisoned because of breaking the law regarding butterflies or disturbing their habitats in the UK; and there has been no person fined or imprisoned because of breaking any European Union (EU) law relating to butterflies in the UK. There have not been enough

Voluntary codes and practices

cases for there to be a judgement as to whether legislative conservation - protecting butterflies by law - is an effective course of action.

The legislative conservation of places where butterflies live, their habitats, such as in National Nature Reserves (NNRs), Local Nature Reserves (LNRs) and Sites of Special Scientific Interest (SSSIs), which are offered some legal protection, has more of a track record which can be judged. It would be convenient if all butterflies lived in protected areas, but they are widely dispersed amongst protected and un-protected areas which are impossible to maintain simultaneously. Butterflies have become extinct locally and nationally in the last few years mainly because of habitat loss but there has been no prosecution for their decline. That some butterflies have become extinct on NNRs and SSSIs which are themselves legally protected has not caused any responsible body or owner to defend itself.

There has thus been no greater stimulus for the voluntary sector to make efforts to conserve the British butterflies, than a shortfall in conservation legislation. The statutory authorities might consider that existing legislation is sufficient to conserve butterflies, but evidence from habitat loss clearly shows it is not. The voluntary sector (as well as the Joint Committee for the Conservation of Invertebrates - the JCCBI) responded by issuing codes of conduct and good practice so that amateurs (and professionals) conform to a well thought out set of guidelines - based on current beliefs; the beliefs, however, do change and mellow with time.

The proliferation of local books on butterflies in the last decade or so, has meant that there are uncontrolled and inconsistent recommendations about how moderate collecting is tolerated county by county. A collector moving across several counties will not know about the county codes and recommendations without first reading the conservation sections of about 20 books. How will he/she know that it is recommended not to take a particular species in one county, say Dorset, but that it can be taken in Cornwall. Collectors will remain collectors and not be too mindful of anyone's recommendations or codes of practice. Laws do exist, however, to restrict commercial trading of some 22 native British butterfly species (Chapter 10).

When the Committee for the Protection of British Lepidoptera was established in 1925, the Victorian and Edwardian passion for collecting had not been finally extinguished. It is interesting that the Minutes of the 25 September of that year refer to having meetings in 'the collecting season'. Over the next 20 years the stance on collecting taken by the Committee mellowed. They presided over the depredations of the commercial collectors who trampled the cliffs of the West Country and took as many prizes as they could muster, and seemed rather helpless in diminishing the ardour of these collectors and their assistants. In the first few years the Committee dealt with all sorts of butterfly conservation issues, and Lord Rothschild called for reserves to be set up in Britain to protect rare butterflies.

Voluntary codes and practices

Norman Riley was asked to complain to a Canadian about an advert for butterflies to be collected 'by the thousand' for his commercial 'decorative work'.

As late as 1953 the Committee saw to it that Mr Hulse was 'written to' following an auction of a long series of Large Blues. Conservation was the main aim of the Committee in these early days, but collecting was a major source of embarrassment. Many members were collectors themselves.

The Committee became more and more vigilant in preserving a place for the butterfly in the British countryside. They published a list of 15 lepidoptera which needed protection, the four butterflies being the Swallowtail, the Glanville Fritillary, the Heath Fritillary and the Large Blue. In 1946 the Amateur Entomologists' Society (AES) were in trouble for wishing to go to print with details of a 'Cinxia hunt' on the Isle of Wight, and Mr Newman had to be written to in March 1951 to stop an advert blatantly asking for certain fritillaries to be collected for the Festival of Britain. In the same year, the Committee were in contact with the National Trust regarding their bye-laws on the taking of insects, and these were distributed with the second annual report of the British Co-ordinating Committee for Nature Conservation.

In a particularly poignant letter to Edlesten, Frank Labouchere wrote on 3 November 1942 that he would rather like the strip of coast between Millook and just beyond The Dizzard to be created a reserve. 'A few years of non-collecting and attention to burning the gorse would soon result in the insect becoming much more plentiful, where permits could be given to ?serious collectors to take a few. At present dealers from all over the country go there and take every insect they come across' (Labouchere, 1942).

Collectors still descend on localities to collect the life stages of certain butterflies, often simply to breed long series so that there is a greater chance of finding varieties. Some entomological journals carry photographs of varieties bred from stock , the stock having in some cases come direct from the wild. In the case of species such as the Duke of Burgundy Fritillary or the Brown Hairstreak localities are descended upon to collect the ova. In the case of the Duke of Burgundy whole fields of Primroses or Cowslips are stripped of all leaves that carry eggs, and in Brown hairstreak areas breeders armed with secateurs cut wands of Hawthorn. On a recent occasion when Berger's Clouded Yellow bred in Portland (Dorset) collectors from all over the country descended on the locality to obtain stock, such is the persistence of this form of lepidopterist (1).

It is therefore vital in the conservation interest of these particular species predated upon by lepidopterists that localities are always kept confidential. However, good localities still become well known and still suffer attack from collectors. This collecting mentality will probably never be stamped out; it is a hangover from the Victorian mania for collecting sports and varieties. Perhaps certain sensitive sites, where the rarest species occur, will have to be policed, as

Voluntary codes and practices

has been done with the Large Blue (from the 1920s to the present day), finances permitting. Disposal of the extra bred stock is another feature that has to be controlled if possible since disposal of extra stock in the wrong place, and with inbred and perhaps weaker stock being released to back-cross with hardy stock, is not in the interests of the insect concerned.

Nothing quite riles the entomologists' hackles than a debate on collecting. It is a very important and fundamental principle which dominates the field, and there is a very obvious dichotomy of thought between amateurs and the few professional lepidopterists. Butterfly Conservation are very much against any form of collecting, but they are not an anti-collecting society, whilst academics and field workers argue quite strongly that there is no evidence that collecting actually adversely affects populations, although they admit there have been a few permanent blimps of extinction. Collecting is clearly dangerous to declining populations. The argument is always given by entomologists that it is essential to collect in order to find out, and some would argue that an entomologist has to go through a collecting phase when young, to become competent in the field. However, the grass-roots butterfly enthusiasts are often completely aghast at any form of collecting.

That so few UK butterflies are protected legally means in reality it is legal to collect 56 out of the 59 species of butterfly in the UK. However, respect of butterfly habitats must be considered before any legal right to collect can be considered. Many butterflies occur in nature reserves owned or managed by voluntary bodies, or occur on municipal land over which there may be voluntary codes or bye-laws which restrict or ban any form of collecting. The legal right to collect must not be assumed anywhere. Furthermore, there is a Nature Reserves Code of Conduct which makes it illegal to collect any insect on any nature reserve in Britain. This is strong stuff, but it is a code with no legal basis. There are also codes of conduct whilst visiting SSSIs, as well as the Countryside Code and the Photographer's Code to consider. Bye-laws have a legal basis and must be strictly adhered to. The National Trust as a leading conservation group have published their own codes with respect to insect collecting, and it can be assumed that any form of collecting is strictly prohibited unless a licence is obtained. The Ministry of Defence also issue permits to enter their properties for *bone fide* research involving collection.

The butterfly professionals who frequently make comments on collecting are never catagoric about the effects of collecting; their texts are full of intimations and full use of the word 'probably' (cf. Morris, Thomas (both numerous occasions), Vane-Wright, 1977. The usual turn of phrase is that examination of the facts does not show anything untoward. Lack of evidence is not evidence that collecting does not cause negative impact. There is in fact more evidence that collecting is dangerous, than that it is not. The often cited comment that butterflies have even

Voluntary codes and practices

become extinct in protected areas, such as NNRs, is presented presumably as evidence that even in the hands of 'Responsible bodies' other factors operate. The fact that butterflies do become extinct in NNRs is more a reflection of our overall ignorance of each butterfly's tolerances, or our ineptitude, or problems relating to lack of funds, since the ground-work on the autecological aspects of the majority of British species has never been done.

There is another facet to the 'moderate collecting is OK' principle proffered by the professionals. These people work at institutions that are based almost entirely on collecting as a *modus operandi* of their entire existence. Many are themselves collectors, and this includes Keepers of Entomology up until recent years. The collecting spirit will obviously take a long time to shake off, since it has been an integral part of the study of butterflies, from before the Aurelians, right up to the present day.

The fascinating, if not bizarre, historical account of Victorian collectors moving from one newly discovered site to another, after pillaging it and taking away their butterfly trophies. The 'moving-on' syndrome of professional collectors is likely to have a significant ecological impact on any species and is guaranteed not to be sustainable, If the population rallied after the collectors' rampages, as it sometimes did as we are told by the 'defenders of collectors lobby', then why are they not there today? Habitat loss and habitat change and degradation could not have attributed exclusively caused to all the extinctions.

That collecting has a negative impact on small populations is unquestionable. It may well have been a major factor in the demise of the Large Blue. Although the comments put forward that colonies could withstand heavy collecting is entirely plausible for large populations, but not presumably every year; cumulative effects must have operated, and presumably large colonies became small colonies eventually.

The collecting debate has continued, sometimes with bitter comment, resentment and internecine rivalry brought about through bad communication between conservation bodies; this resulted in the loss of at least one Heath Fritillary colony in Devon (see Pyle 1981). Pyle (2) summed up the situation by saying the Joint Committee for the Conservation of British Insects (as it was then) issued a *Code of Collecting* (Appendix 2) which espoused a moderate collecting policy, and that it seemed to him to be a sound collecting policy which appeared to work. The constraints of conservation and collecting, seemingly not good partners, have to be reconciled. If there is room for a criticism it is that the JCCBI policy is a collecting policy, not a conservation policy based on non-collection. Jeremy Thomas believes that nothing more can be added to the JCCBI code (Thomas, 1994), but it could have a conservation facelift with a different emphasis. For serious lepidopterists and conservationists seen wielding a net in the field (whether authorized or not) there is a very real possibility that some people will

Voluntary codes and practices

become hostile, thinking that indiscriminate collecting is being done. In effect there are probably only a few rogue collectors, but one has only to have a few of these to eliminate populations that are already at low numbers. When collecting was at its height the impact was not as great as it would be today, since butterfly populations were generally much stronger.

If the issues of collecting *per se* were not enough in general terms to be debated by all lepidopterists, the aims of the JCCBI were to try to restrict collecting of certain butterfly species thought to be threatened. The trend of naming threatened species generally started earlier this century with the Committee for the Protection of British Butterflies under the Chairmanship of Lord Rothschild. The formation of the Conservation Committee of the Royal Entomological Society of London (CCRESL) in 1925 was as a direct result of the many complaints in the entomological journals of the time about the work of unscrupulous collectors (Southwood, 1965). In 1957 the RESL formulated a long list of insects of many orders which they thought needed to be placed on a 'Confidential list of rare localised British insects' of which there had been nine butterflies. Four of these - the Large Copper, the Silver-studded Blue subspecies *caernensis*, the Brown Argus and the Grayling - were dropped from the 1965 conservation appeal list.

Reports of the mass collection of butterflies were still being received by the Conservation Committee in the mid 1960s, so it was that the RESL issued an appeal for 'insect conservation', a carefully drafted note for restraint by collectors. Six species of butterfly were named, along with 12 species of moth: 'collectors are again earnestly requested to use the utmost restraint in taking any of the species listed below in any of their stages,' Amongst the 19 species of lepidoptera were six species of butterfly: Swallowtail, Glanville Fritillary, Heath Fritillary, Large Blue, Adonis Blue and Black Hairstreak. Only the Large Blues were collectors asked to 'abstain' from collecting, since it was 'evidently on the verge of extinction in England'. Southwood reported that only 80 specimens of Large Blues had been reported for 1963. As Southwood said (Southwood, 1965) the 1925 Committee was set up to guard against unscrupulous collectors. The pressure was still on against collectors 40 years later.

Collecting of the Large Blue was forbidden by the JCCBI in 1972, and collection of the Large Copper at Woodwalton Fen was also forbidden by the Nature Conservancy (JCCBI, 1972). This new publication of a long list of lepidoptera for which 'The Committee recommends that the other species (apart from the large blue) and subspecies listed should be collected with the greatest restraint and suggests that a pair of specimens is sufficient' brought into the fold more species than had otherwise been regarded as threatened. A lot of people would be aghast at the suggestion of only taking two specimens of some of these species in the mid-1990s, in fact Butterfly Conservation would regard any form of collecting 'as completely unnecessary and undesirable'. **(1)**

Voluntary codes and practices

In 1972 the JCCBI published a list of 15 species and races for which restraint should be exercised in collection, i.e. no more than two specimens was deemed sufficient, with absolutely no collection of the Large Blue or the Large Copper (JCCBI, 1992). The species included the Chequered Skipper, Large Heath (in England only), Scotch Argus (in England only), Grayling (*Hipparchia semele thyone*), Silver-spotted Skipper, Adonis Blue, Large Blue, Heath Fritillary, Glanville Fritillary, Large Tortoiseshell, Swallowtail, Silver-studded Blue (*caenensis* and *masseyi*), Black Hairstreak and Lulworth Skipper. The problem about putting a limit of two specimens of each species is that small populations would not stretch to all members of Butterfly Conservation as an example. But wouldn't it be nice to have 20,000 individuals of the Large Tortoiseshell in England anyway?

RE-INTRODUCTIONS

The noted lepidopterist L. Hugh Newman was never against re-introductions so long as they were well managed by the proper authorities, and he held the work of the Committee for the Protection of British Insects in high esteem. It was Newman who executed a unique experiment on re-introduction using the Black-veined White, *Aporia crataegi*, at Chartwell, Winston Churchill's home. This was done despite some local bureaucracy, since the Black-veined White had been on the Ministry of Agriculture's list of forbidden pests. Fortunately, Sir John Fryer, who was chief entomologist to the Ministry of Agriculture and Fisheries (Fryer was also on the Committee for the Protection of British Insects) cleared the species for approval and experiments were made to establish the butterfly at Chartwell. The pest butterfly had by then been taken off the pest list. Churchill wondered about the possible ravages the larvae might have had on his fruit trees, but all was well. The butterflies did well only for that year of introduction. The conservation point is that the butterfly could have prospered beyond everyone's beliefs and become a pest in the Garden of England amongst the highest concentration of fruit trees in Britain. It was thought a considered risk worth taking, and it met with limited success. It is not generally regarded that the Black-veined White was ever a serious pest in England. There have been attempts at re-introducing the butterfly again, but nothing substantial has happened.

The Black-veined White reintroduction to Chartwell in Kent failed, did as the well-documented introduction of the Map Butterfly, *Araschnia levana* to the Forest of Dean; but this latter species died out for another reason, as Hugh Newman reminds us: a local collector, A.B. Farn, was so upset he successfully exterminated the colony. To introduce an exotic or alien species into the British countryside is not well approved of, and is against all the codes of practice that

Voluntary codes and practices

various bodies have sought to clarify. Such an introduction today might also stimulate a similar response.

The release of butterflies back to their former habitats, and to new ones, has always been a source of debate. Lepidopterists are forever collecting material from localities, and releasing their excess breeding stock perhaps at a different site. The countryside has been cluttered up with an assortment of introduced material of unproven provenance. Sorting out the proper localities and requirements of butterflies from the available evidence of distribution today has taxed the minds of distribution ecologists. Some order and control to the mass of releases has always been extolled, but is not likely to be widely accepted. Covert releases are thus always to be expected.

SPNRs INTRODUCTION CODE, 1968

It had been an objective of the Committee for the Protection of British Lepidoptera to introduce butterflies into their former haunts since the early part of the twentieth century, and much of this is documented in the Committee's Minutes, especially to do with the Large Copper. It was not until the late 1960s, however, that the Society for the Promotion of Nature Reserves (SPNR) published its A Policy on Introductions to Nature Reserves (Appendix 3). The SPNR code built substantially on a set of introduction policies published by the International Union for Conservation of Nature and Natural Resources in *Problems in Species' Introductions* IUCN 1968, Bulletin N.S.2., 70-71) an extract of which can be seen in Working Group (1979).

The SPNR code was much concerned with species conservation and reserve management. Regarding the criteria for introductions to reserve areas the SPNR set the scene by saying that if a species disappeared from a reserve, or adjacent to one, or from a place that was not a reserve (i.e. anywhere) then there were grounds for introducing the species. Surprisingly there were also measures for removing species should they become dominant in a reserve area; thankfully butterflies have not yet reached epidemic proportions like plants have.

It was during this time that the AES published their Conservation Responsibility code for those involved with trading in living and dead material (Appendix 4). There was concern in some quarters that the trade in insects, particularly those from overseas, was getting a little out of hand, so these measures and guidelines were published. It was suggested by Pyle (1981) that the JCCBI produced this code, but I gather from Mike Morris that this was not so.

WORKING GROUPS INTRODUCTION CODE, 1979

That uncontrolled introductions of many groups of animals and other organisms were at the heart of the debate of the authorities was evident in 1979 when the UK Committee for International Nature Conservation published their results after a round-table meeting of some 33 voluntary conservation bodies; this included the Society for the Promotion of Nature Conservation (Working Group on Introductions of the UK Committee for International Nature Conservation, 1979). Unfortunately neither the RESL nor the then British Butterfly Conservation Society was represented at the meeting (which did not represent formal government policy, although eight government bodies were represented) but the bodies that were present did discuss some butterfly introductions and extinctions, most notably the Large Copper becoming extinct in 1927 (when they were not concerned with Coypus in England or Parma Wallabies in New Zealand.) They also stated that there had been 'no established introduced species in Britain, despite the widespread breeding of very many exotic species'.

That one particular introduction of an organism, that of the myxomatosis virus, had a profound environmental impact on butterfly ecology and viability, was not lessened by the fact that up to 1979 there were already in place 28 separate Acts and Orders of Parliament which relate to introductions in the UK (see their Chronological List). It might be a pleasant exercise for someone interested in butterfly legislation to scour these various Acts and Orders to find suitable tenets of legislation which would afford butterflies some legal protection. That the law has not previously been exercised for the benefit of butterflies, especially as so many lay (even legal minded) people do not appreciate that butterflies (and moths) are indeed 'animals', may well open the door for butterfly conservation and preservation in the future.

Two scientists who knocked some quantitative sense out of this abyss of uncontrolled releases were Matthew Oates and Martin Warren who published their report on introductions in 1990. It is highly recommended. A lot of attention has been given to the ethics of introductions and establishments, and to this end the JCCBI published their own exhaustive code which is reproduced here in Appendix 5.

BUTTERFLY CONSERVATION'S RELEASES CODE, 1990

Butterfly Conservation published their first code of practice for butterfly releases when they were the British Butterfly Conservation Society. It has been up-dated regularly, with few amendments (Appendix 6). The code rather assumes that

Voluntary codes and practices

unrestricted introductions will be taking place, and therefore its points should be adhered too as much as possible. The problem with all codes is that they hardly ever get policed or monitored by the statutory authorities.

Butterfly Conservation have a very detailed conservation strategy (the summary of which is shown in Appendix 7), and one section relates specifically to Repopulation Policy. They are committed to supporting 'properly-planned and monitored re-establishment projects'....'where they are considered by the Conservation Committee to be desirable and feasible'. Butterfly Conservation are investigating the possibility of bringing back the Chequered Skipper to England, and working with other bodies to re-establish the Large Blue, the Large Copper and the Swallowtail butterflies. Butterfly Conservation actively discourage unplanned releases of lepidoptera and ask all their members to obey the existing Code of Practice.

Notes

1. I am grateful to Patrick Roper of Butterfly Conservation for bringing me up to date on the callousness of present-day collectors and breeders.

2. Dr. Robert Michael Pyle is an American who has studied comparative butterfly conservation theory. He prepared his doctorate on The eco-geographic basis for Lepidoptera conservation at Yale University in 1976, and during the late 1970s worked at the Species Conservation Monitoring Unit of the International Union for the Conservation of Nature at Cambridge, England.

CHAPTER TWELVE
NATURE RESERVES

'The mere acquisition of nature reserve for rare insect species is but one step in insuring their protection and long-term survival.' Thomas, 1981b

'Conservation management is economically marginal' McLean, 1993

If we are not careful many of our butterflies will be confined just to nature reserves in Britain. This is the view of some, but it is a very unlikely scenario since British butterflies have great adaptations to living in a landscape moulded by people. Many of the commoner species will always thrive in parks, gardens and wastelands alongside man. But for the specialist butterflies with rather specialized habitat requirements their choice of habitats will be much more

Nature Reserves

restricted. They will have been wiped out from many of the interesting habitats and will not be represented in the vast vacuums of agricultural land stripped of all original or semi-natural habitats. Nature reserve acquisition is not enough, as Jeremy Thomas (1981b) reminds us, but what alternatives are there? Nature reserve acquisition is a pathway that conservationists have followed in the absence of other proven methods, but it does not ensure the conservation of species. The 'bottom line' is that butterflies are not yet big business and do not turn the turnstiles. It is unlikely that they ever will. The conservation of Britain's butterflies will have to continue as best it can dove-tailed into the agricultural, industrial and urban environment that confronts it, even if conserving butterflies is sometimes regarded as marginal.

The system of nature reserves in Britain is pretty well defined. There are statutory nature reserves, such as NNRs, LNRs and FNRs, and there are the many more run by voluntary bodies, otherwise called Non-Governmental Organizations (NGOs). This chapter is really about the general range of butterflies in nature reserves - so that one can consider their relative conservation within these areas - but it also pays special attention to NGO nature reserves for butterflies, especially those owned or managed by the Wildlife Trusts and Butterfly Conservation.

There are 2,000 reserves owned or managed by The Wildlife Trusts, occupying some 52,000 ha, and 118 reserves owned by the RSPB which cover a much larger area of 74,700 ha. Not that bird reserves have primarily been managed in best interest of butterflies in the past, but there is definitely a policy swing within the RSPB to cater a little more for butterflies and other insects in Britain, after all, butterflies are food for some birds, apart from looking good on heathland and grassland reserves.

With a big stretch of the imagination one could almost think of SSSI as nature reserves (it would be really pretty good for wildlife if they were, and had their range of usual restrictions on collecting and the like) but, although the SSSIs represent the cream of all the habitats in Britain they cannot be relied upon to conserve Britain's butterflies as nature reserves. The latest count puts them at 3,749 in England, covering 861,341 ha, representing 6% of England's total and about 18,000 people own or occupy these sites (EN, 1993c: Autumn, 1993 statistics). There is negative, and significant environmental impact occurring on SSSIs each year, some of which is bound to affect butterflies. What is most dangerous for butterflies on SSSIs is what happens around the periphery of the SSSIs (such as habitat degradation and loss) since these marginal areas (especially if they are woodland margins) are especially rich in butterflies, and are therefore highly susceptible to change. The same troublesome impacts also affect nature reserves, it is just that SSSI areas are often more important localities than other blocks of countryside. In many undisturbed and neglected woodlands the marginal areas are often the best places for butterflies.

National Nature Reserves (NNRs) are a little safer and more permanent than the highly vulnerable SSSIs, and butterflies fare much better in these bone fide nature reserves. NNRs tend to be large and all are open to the public. One NNR specifically managed for butterflies is Glapthorn Cow Pastures in Northamptonshire (Northamptonshire Trust, 1994 pers. comm.), but this is not the normal situation on NNRs which have to conserve a wide range of habitats and flora and fauna. All NNRs are now SSSIs so they have all the added protection that SSSI status confers. So how have butterflies fared on NNRs?

Results from the latest survey of information from NNR managers undertaken by Ian McLean of English Nature (McLean *et al.* 1993) indicated that the management of butterflies was taking place on 81 NNRs and that there were 53 with Butterfly Monitoring Survey (BMS) transects operating on them. There were a further 28 butterfly transects that were not part of the BMS scheme on NNRs. The survey also showed that 51 of Britain's 59 species were breeding on NNRs. This is quite a satisfying result. Maybe most of Britain butterflies will be safe under NNR protection.

As for the health of breeding butterflies on NNRs, butterflies appeared to be doing well. The following were the top ten most abundant breeders, in this order: Small Tortoiseshell, Green-veined White, Meadow Brown, Small Heath, Common Blue, Peacock, Small Copper, Orange Tip, Speckled Wood and Wall. Other aspects of this survey showed that the Swallowtail was doing very well in the Hickling Broad area because of the high level of management put in, and, in contrast, that the Mountain ringlet was declining in the Highlands. In Cornwall, the Duke of Cornwall was able to have some recently planted conifers removed to make way for the expanding populations of Heath Fritillary; this turned out to be a great success for the butterfly on that site. Perhaps the most worrying assessment that McLean made was that 'conservation management is economically marginal'. Not many people can conserve butterflies and make a good profit out of a woodland. Butterflies do not bring in cash, yet, and do not figure high on the list of species that pull tourists on the trail of eco-tourism.

One of the benefits of having reserves for wildlife, even for butterflies in particular, is that their presence can be used to strengthen the case for conservation or preservation in the face of development. In Cornwall, for instance one reserve called Loggans Moor, managed by the Cornwall Trust for Nature Conservation, was finally purchased thus saving it from development.

For most enthusiasts studying butterflies in the field, their point of focus is often their local wildlife trust nature reserve. With this in mind, I thought it useful to compile a list of sites, county by county, based around those that have a legal right of access. With the butterflies' welfare at stake, all 47 Trusts

Nature Reserves

belonging to The Wildlife Trusts were contacted by letter for information on nature reserves and good butterfly sites in their county which had a legal right of access and which could be published in this book; Appendix 8 gives the results from the 23 County Trusts that responded.

LOCAL RESERVES AND CONSERVATION

The boom in butterfly books, county by county, has been a great asset in the understanding of the needs of each species locally. There is now much more circumstantial information bearing on butterfly conservation coming from groups of enthusiasts locally than from the few professional butterfly experts. The swell of butterfly enthusiasts has helped to super-charge this enthusiasm at a local level. The result has been a proliferation of local butterfly books. They vary in the amount devoted to conservation from virtually nothing such as for Kent (Philp, 1993), London (Plant, 1987) Durham and Northumberland (Dunn & Parrack, 1986) and Brecon (Sankey-Barker, et al 1978) to perhaps the most comprehensive, for instance, Yorkshire (Sutton & Beaumont, 1989). Those for Sussex (Pratt, 1981) and for Suffolk (Mendel & Piotrowski, 1986) have some useful pieces of local conservation information. It is encouraging to see sections on conservation appearing in books on the butterflies of various counties today, even though they are often only cursory.

The benefit of local records, backed by the fast increasing ranks of Butterfly Conservation, is that there are more people involved, who know their local habitats intimately and who can report back up the information corridor to the local nature conservation groups. They can also monitor situations on the ground, month by month and year by year better than anyone else. Their results are immensely valuable, especially on a comparative basis, and continue to be so especially on the useful BMS scheme.

There is nothing like a local butterfly nature reserve to whip up the enthusiasm of butterfly-lovers, and it is heartening that Butterfly Conservation now have a substantial network of butterfly reserves throughout much of England (Table 12.1). Some are owned by BC, whilst the remainder are managed for butterflies by a variety of means. The names of some of BC's nature reserves are very personal, reflecting the generosity of those who have given the reserve to BC: for instance the Bill Smyllie reserve, the Mike Brown reserve and the Tony Steele reserve. Butterfly Conservation have struck up a sound relationship with the Forestry Commission and have worked closely with the establishment of Forest Nature Reserves, Forest Enterprise Reserves and Conservation areas. The future of butterflies on Butterfly Conservation's nature reserves seems to be one of increasing fortunes. Intending visitors to BC nature

Nature Reserves

reserves and other sites, should realise that many of the sites do not have public access and that they should therefore check with BC beforehand for permission and access details.

Table 12.1 Nature Reserves Owned or Leased by Butterfly Conservation

Name, Status, County, Grid Reference, Size, Tenure

Bentley Station Meadows, SSSI, Hampshire (SU 794429) 4.2ha owned
Bill Smyllie (Prestbury Hill) SSSI (pt), Gloucs (SO993245) 32.0ha owned
Broadcroft Quarry, SSSI, Dorset (SY686709) 7.5ha leased
Catfield Fen, SSSI, Norfolk (TG368214) 23.8ha owned
Little Breach, SSSI, Somerset (ST115154) 0.4ha owned
Magdalen Hill Down, Hampshire (SU502292) 9.5ha leased
Monkwood, SSSI, Worcestershire (SO804607) 61.4ha owned
Narborough Railway Line, Norfolk (TF750118) 6.0ha agreement
Park Corner Heath, SSSI, East Sussex (YQ516147) 4.0ha owned
Snakeholme Pit, Lincolnshire (TF120720) 1.0ha agreement
Southrey Wood, Lincolnshire (TF128682) 9.0ha agreement
Stoke Camp, Somerset (ST491510) 11.0ha owned
Thurlbear Quarrylands, SSSI, Somerset (ST272209) 8.3ha leased
Trench Wood, Worcestershire (SO929588) 42.0 agreement
Yew Hill, Hampshire (SU455265) 3.8ha leased

Other Nature Reserves with substantial BC involvement

Ashton Upthorpe, Oxfordshire (private owner)
Bernwood Forest, Buckinghamshire, including Waterperry Wood, Oxfordshire
Blacklands Copse, Buckinghamshire (private owner)
Bodham butterfly oasis, Norfolk (rented from North Norfolk District Council)
Buckland Wood, Blackdown Hills, Somerset (Forest Enterprise Butterfly reserve)
Cander Moss (Scottish Wildlife Trust reserve)
Church Wood, Blean (RSPB reserve)
Coombe Hill, Buckinghamshire (NT property)
Doddershall Wood, Buckinghamshire (private owner)
Dowles Valley (Wyre Forest) Shropshire (informal management agreement with EN)
Finemere Wood, Buckinghamshire (BBNOT reserve)
Great Breach Wood (STNC reserve)

Table 12.1 cont..

Nature Reserves

Lydford Gorge, Devon
Lydlinch Common, Dorset (formal management agreement)
Mount Faney & Bow Green, Somerset (Forest Enterprise Reserve, FER)
Oak Plantation, Staple Common, Somerset (FER, informal management relationship between private individual and FC covering ride at north end
Northwood Hill, Kent (RSPB reserve)
Osgodby Moor, Lincolnshire (Forestry Commission Conservation Area)
Fittington Hill, County Durham (private owner, potential BC reserve)
Prestwood Picnic Site, Buckinghamshire (neither owned nor leased by BC)
Pulpit Hill, Buckinghamshire (National Trust property)
Rosyclose Copse, Somerset (Forest Enterprise Butterfly reserve)
Ruttersleigh Common, Somerset (Forest Enterprise Butterfly reserve)
Shipley Railway Station Meadow, Bradford, West Yorkshire (British Rail)
Snakeholme Pit, Lincolnshire (formal management agreement with Anglia Water, not leased)
Southrey Wood, Lincolnshire (formal management agreement with the FC, not leased)
Stoke Camp, Mendips (purchased being pursued by BC, February 1994)
Tony Steele reserve, also known as Preston Hill SSSI (Kent County Council nature reserve)
Trosley County Park SSSI, Kent (Kent County Council Country Park)
Twyford Butterfly Sanctuary, Lincolnshire (owned by British Steel, managed by FC; BC have informal management arrangement
White Hill, Shoreham, Kent - Mike Brown reserve (owned by Dunsany estate, BC have informal management agreement
Yeesden Wood/Bank, Buckinghamshire (private owner, potential BC reserve)

Sources: Compiled by Paul Waring up to March 1994 for Butterfly Conservation, and from Butterfly Conservation Annual Review 1994. N.B. There are many other sites where Branches help with management tasks, details in Branch newsletters.

Not only are BC establishing a sound base of nature reserves to protect the widest range of butterfly species in Britain, but they also mount highly publicized 'health check trips' to various butterfly localities. In 1993 they focused media attention on some of the finest reserves for butterflies including some of the most endangered species (Table 12.2) and in 1994 their 'health checks' included woodland butterflies (Table 12.3).

Nature Reserves

Table 12.2
Butterfly localities visited by Butterfly Conservation's Butterfly Check '93

This is the list of 25 butterfly sites that the Butterfly Conservation team (sponsored by Land Rover) checked out during the 1993 season. The purpose of the visit was to check the situation with regard to Britain's rarer butterflies, especially those protected by the Wildlife & Countyside Act 1981. The visits started on 6th May in Essex finishing in West Sussex on the 26 August, in the order presented in this chart. The butterflies (or their life stages) mentioned in the third column are those which members of the team saw, or should have seen, on the site visits. There were three sites which did not come up to expectation - no Duke of Burgundy in Yorkshire on 20 May, no High brown fritillaries on Exmoor on 6 July and no Brimstones were seen at Onchan on the Isle of Man on 10th August. Other comments of a conservation nature are included in column three. All data comes from a Press Release from Butterfly Conservation.

Little Baddow Heath, Essex. Pearl-bordered fritillaries, Grizzled skipper re-established as part of the Essex Butterfly Action Plan

Brown Hill Quarry, Holwell, East Midlands, Dingy skipper

Pexton Bank Forest, Yorkshire. Duke of Burgundy fritillary

Pyle Industrial Estate, Glamorgan. South Wales, Marsh fritillary

Monkwood, Droitwich, W. Midlands. Increasing population of wood whites

Strontian, Scottish Highlands. Chequered skipper

Inishargy Bog, Ards Peninsula, N. Ireland. Marsh fritillary

Old Totternhoe Quarry, Dunstable, Bedfordshire. Small blues

Compton Chine, Isle of Wight. Glanville fritillary

Catfield Fen, Norfolk. Swallowtail

Crowl Waste, Isle of Axholme, Lincolnshire. Large heath, under threat here due to Fisons mining surround land for peat.

Bernwood Forest, Buckinghamshire. Black hairstreak in Forest Enterprise Reserve.

Blean Woods, Canterbury, Kent. Heath fritillary

Secret site, Devon. Large blues

Holnicote Estate, Exmoor, Somerset. High Brown fritillary

Broadcroft Quarry, Portland, Dorset. Silver-studded blue

Botany Bay Wood, Chiddingfold, Surrey. Purple emperor

Durlstone County Park, Dorset, Lulworth skipper

Bill Smyllie Reserve, Gloucestershire. Chalkhill blue

Table 12.2 cont..

Nature Reserves

Feoch Meadows, Ayreshire. Scotch argus doing well following sympathetic management
Onchan, Isle of Man. Brimstone
Alfriston, nr. Eastbourne, East Sussex. Silver-spotted skipper
Magdalen Hill Down, Winchester, Hampshire. Brown argus
Ipplepen, nr. Newton Abbot, Devon. Brown hairstreak
Mill Hill LNR, Shoreham, West Sussex. Adonis blue

Table 12.3 Butterfly Conservation's Land Rover's Woodland Campaign Roadshow, 1994 (listed chronologically from May 25th)

Woodlands visited	Target species observed
Heath House Woods, Hampshire	Pearl-bordered Fritillary
Pickett Wood, Wiltshire	Grizzled Skipper
Monkwood, Worcestershire	Wood White
Glasdrum NNR, Oban, Scotland	Chequered Skipper
Dalby Forest, Yorkshire	Duke of Burgundy Fritillary
Luckett Wood, Cornwall	Small Pearl-bordered Fritillary
Monkswood NNR, Cambridgeshire	Black Hairstreaks
East Blean Woods, Kent	Heath Fritillary
Gait Barrows NNR, Lancashire	High Brown Fritillary
Bernwood Forest, Oxfordshire	Purple Emperor
Bardney Forest, Lincolnshire	White Admiral
Glen Arm, County Armagh	Silver-washed Fritillary
Holm Wood Common, Dorking	Brown Hairstreak

OTHER NATURE RESERVES

There are numerous nature reserves run by various voluntary conservation bodies, including the county branches of the Wildlife Trusts Partnership, which harbour butterflies. Full details are available from your local offices and libraries. The sites listed in Appendix 8 are all those which have been mentioned by each of the local wildlife trusts following a request for information. Each is open to the public, though some are only accessible only via a footpath or bridlepath meaning that not all of the reserves are open access. Permits are not required to visit any of the sites. To find these sites it is prudent to purchase the appropriate Trust booklet which gives full details of each site, map reference, ecological details, species lists, and often, a small map of the site. Many of the Trusts have been involved with production of books of the butterflies of their county. Wildlife Trusts often have many other nature reserves (not listed below) which are rich in butterflies (perhaps richer than some listed below), but these are open by permit only.

Butterflies on Royal Society for the Protection of Birds (RSPB) reserves has been assessed by Cadbury (1990). Forty-six species were suspected to have bred on RSPB reserves in Britain and Ireland during the 1980s and this included three vulnerable RDB species (Swallowtail, High Brown Fritillary and Heath Fritillary) and nine nationally scarce species. Some habitats owned by RSPB have been managed both for birds and butterflies with good results, though there has been some local extinctions and fluctuations which is to be expected. Calcareous grassland is poorly represented in RSPB hands in southern Britain resulting in a consequent lack of species normally found there.

Britain probably leads the world in the number of designated reserves for wildlife; there is probably plenty of opportunity to go out in your local neighbourhood and discover butterflies on and off nature reserves.

APPENDICES

APPENDIX 1: Species action plan for the Large Blue Butterfly, Maculinea arion

Formerly listed as a globally threatened species; Schedule 5, Wildlife and Countryside Act

Current status
Formerly extinct but now successfully re-established at four sites.

Main threats
Loss of habitat, coupled with lack of grazing or other appropriate management.

Objectives
1. To consolidate the re-established populations at four sites in England.
2. To re-establish populations at a further six former sites in southern England.

Broad Policies
1. To increase the area of suitable habitat at the four sites of re-establishment by active management, control of grazing and advice to adjacent owners or by land purchase.
2. To re-establish and maintain grazed and scrub-free suitable habitat at six former sites, where necessary by advice and management agreements with owners or by land purchase.
3. As suitable habitat becomes established, conduct further translocations to the six former sites.

Proposed Actions
1. Policy and legislative
   Promote the beneficial management of land for the Large Blue in areas adjacent to the existing and future sites of re-establishment by including appropriate management prescriptions in management contracts for Environmentally Sensitive Areas (ESAs), long-term set-aside and other agri-environment regulation schemes.

Appendix 1 cont..

Appendices

2. Site safeguard

a) SSSI designation to be extended to all target re-establishment sites and SSSI citations to be amended to include the large blue once re-establishment has taken place.

b) Encourage positive management of SSSI through management agreements.

c) Oppose any development proposals threatening target and other potential translocation sites.

3. Land Acquisition and Reserve Management

a) As opportunity arises EN or Wildlife Trusts should seek to acquire key areas within each target site as preferable option to management agreements.

b) Manage all holdings and adjacent peripheral areas, subject to agreements, to ensure suitable sward and scrub-free habitat.

c) Where necessary plant wild thyme in potential re-establishment sites.

4. Species management and protection

Low key wardening is essential at all existing and future re-establishment sites to prevent illegal collecting or incidental disturbance.

5. Advisory

Produce simple advisory leaflet on how to manage land for the large blue for circulation to target landowners and managers in suitable areas adjacent to target sites.

6. International

Use experience gained from conserving this species in the UK to advise other countries about how to save populations of this and other four species of large blue in Europe.

Reproduced with kind permission of the RSPB. from *Biodiversity Challenge*, 1994 (Royal Society for the Protection of Birds, 1994).

Appendices

APPENDIX 2: A Code for Insect Collecting, 1972

This Committee believes that with the ever-increasing loss of habitats resulting from forestry, agriculture, and industrial, urban and recreational development, the point has been reached where a code for collecting should be considered in the interests of conservation of the British insect fauna, particularly macrolepidoptera. The Committee considers that in many areas this loss has gone so far that collecting, which at one time would have had a trivial effect, could now affect the survival in them of one or more species if continued without restraint.

The Committee also believes that by subscribing to a code of collecting, entomologists will show themselves to be a concerned and responsible body of naturalists who have a positive contribution to make to the cause of conservation. It asks all entomologists to accept the following Code in principle and try to observe it in practice.

1. Collection - General
1.1   No more specimens than are strictly required for any purpose should be killed.
1.2   Readily identified insects should not be killed if the object is to 'look them over' for aberrations or other purposes: insects should be examined when alive and then released where they were captured.
1.3   The same species should not be taken in numbers year after year from the same locality.
1.4   Supposed or actual predators and parasites of insects should not be destroyed.
1.5   When collecting leaf-mines, galls and seed heads never collect all that can be found; leave as many as possible to allow the population to recover.
1.6   Consideration should be given to photography as an alternative to collecting, particularly in the case of butterflies.
1.7   Specimens for exchange, or disposal to other collectors, should be taken sparingly or not at all.
1.8   For commercial purposes insects should be either bred or obtained from old collections. Insect specimens should not be used for the manufacture of 'jewellery'.

2. Collecting - rare and endangered species
2.1   Specimens of macrolepidoptera listed by this Committee (and published in the entomological journals) should be collected with the greatest restraint. As a guide, the Committee suggests that a pair of specimens is sufficient, but that those specimens in the greatest danger should not be collected at all. The list may be amended from time to time if this proves to be necessary.

Appendix 2 cont..

Appendices

2.2   Specimens of distinct local forms of Macrolepidoptera, particularly butterflies, should likewise be collected with restraint.

2.3   Collectors should attempt to break new ground rather than collect a local or rare species from a well-known and perhaps over-worked locality.

2.4   Previously unknown localities for rare species should be brought to the attention of this Committee, which undertakes to inform other organisations as appropriate and only in the interests of conservation.

3. Collecting - Lights and Light traps
3.1   The 'catch' at light, particularly in a trap, should not be killed casually for subsequent examination.
3.2   Live trapping, for instance in traps filled with egg-tray material, is the preferred method of collecting. Anaesthetics are harmful and should not be used.
3.3   After examination of the catch the insects should be kept in cool, shady conditions and released away from the trap site at dusk. If this is not possible the insects should be released in long grass or other cover and not on lawns or bare surfaces.
3.4   Unwanted insects should not be fed to fish or insectivorous birds and mammals.
3.5   If a trap used for scientific purposes is found to be catching rare or local species unnecessarily it should be re-sited.
3.6   Traps and lights should be sited with care so as not to annoy neighbours or cause confusion.

4. Collecting - Permission and Conditions
4.1   Always seek permission from landowner or occupier when collecting on private land.
4.2   Always comply with any conditions laid down by the granting of permission to collect.
4.3   When collecting on nature reserves, or sites of known interest to conservationists, supply a list of species collected to the appropriate authority.
4.4   When collecting on nature reserves it is particularly important to observe the code suggested in Section 5.

5. Collecting - damage to the environment
5.1   Do as little damage to the environment as possible. Remember the interests of other naturalists; be careful of nesting birds and vegetation, particularly rare plants.
5.2   When 'beating' for lepidopterous larvae or other insects never thrash trees and bushes so that foliage and twigs are removed. A sharp jarring of branches is both less damaging and more effective.

Appendix 2 cont..

Appendices

5.3   Coleopterists and others working dead timber should replace removed bark and worked material to the best of their ability. Not all the dead wood in a locality should be worked.

5.4   Overturned stones and logs should be replaced in their original positions.

5.5   Water weed and moss which has been worked for insects should be replaced in its appropriate habitat. Plant material in litter heaps should be replaced and not scattered about.

5.6   Twigs, small branches and foliage required as foodplants or because they are galled, e.g. by clearwings, should be removed neatly with secateurs or scissors and not broken off.

5.7   'Sugar' should not be applied so that it renders tree-trunks and other vegetation unnecessarily unsightly.

5.8   Exercise particular care when working for rare species, e.g. by searching for larvae rather than beating for them.

5.9   Remember the Country Code!

6. Breeding

6.1   Breeding from a fertilised female or pairing in captivity is preferable to taking a series of specimens in the field.

6.2   Never collect more larvae or other livestock than can be supported by the available supply of foodplant.

6.3   Unwanted insects that have been reared should be released in the original locality, not just anywhere.

6.4   Before attempting to establish new populations or 'reinforce' existing ones please consult this Committee.

Reproduced by kind permission of the Joint Committee for the Conservation of British Invertebrates.

Appendices

APPENDIX 3: Summary of RSNC Policy on introductions.

1. Introductions should be kept to the minimum necessary for the furtherance of the objectives implied in the reasons for reserve establishment. From this it follows that:

(i) only those species typical of the ecosystems represented on the reserve and of that particular region should be introduced;

(ii) individuals for introduction should be taken from populations in the same part of the geographical range as the reserve, except in the case of species close to extinction when other steps may have to be taken;

(iii) populations of introduced species should not be allowed under management to get unnaturally large;

(iv) special attention needs to be paid to species in the higher trophic levels which are particularly vulnerable;

(v) alien species should not generally be introduced to reserves, and aliens already present should be discouraged or exterminated, except where they form part of an economic crop;

(vi) any new habitats created should be of a viable size;

(vii) new habitats created should not fragment existing ones;

(viii) new habitats created should be relevant to those already in existence on the reserve.

2. Exceptions should be made to these general principles where they affect specific research projects.

3. All introductions should be adequately recorded and details sent to the Nature Conservancy's Biological Records Centre.

Reproduced with kind permission of The Wildlife Trusts. Taken from the Policy on Introduction to Nature Reserves. SPNR. Conservation Liaison Committee. Technical Publication No. 2. 1970. This work was a policy document which laid the foundation for further conservation policies evolved with other conservation bodies.

Appendices

APPENDIX 4: A Code for Dealers

At its November meeting, the Council of the AES considered details of "A Code of Conservation Responsibility" agreed by a group of firms dealing in live and dead insects. The Council welcomed this initiative and it was agreed that the code and the names of those persons or firms subscribing to it should be published in the Society's Bulletin. The details are set out below.

Conservation Responsibility
In November 1974 a Meeting of the main Entomological Suppliers was convened in London to agree a code of Conservation Responsibility. The following points form the initial code:
1.) A "Red List" of endangered species from overseas will be kept and any species on the list will not be bought by the Parties to the code (who are named below) or offered for sale. The list will be as brief as possible, and only species relevant to those which are likely to be imported will be included. The Parties to the Code of Conservation Responsibility will regularly review the question of species for the Red List. The species will be confined to those whose need is critical and in full agreement of those concerned. The question has arisen over *Parnassius apollo*
which is declining in some areas, yet is abundant in others. In this case the species may be restricted from defined areas.
The first three species are:-
>   *Troides aeacus kaguya*
>   *Delias hyparete peirene*
>   *Delias aglaia*

All from the island of Taiwan

2.) Specimens of Protected or Endangered British species will not be collected for sale at all. Parties to the code wish to stress that sale of such specimens from old collections will continue and it is by this means that collectors can obtain rarities.

Species that will not be collected for sale as collectors' specimens:-
>   Large Blue (*Maculinea arion*)
>   Chequered Skipper (*Carterocephalus palaemon*)
>   Large Tortoiseshell (*Nymphalis polychloros*)
>   Black Hairstreak (*Strymonidia pruni*)
>   Brown Hairstreak (*Thecla betulae*)
>   Heath Fritillary (*Mellicta athalia*)

Appendix 4 cont..

Appendices

3.) Offered series of wild-caught British Specimens will be refused if they have been caught for the purpose of selling as a stock to suppliers. Genuine duplicated and 'thinnings' from a collection are considered as being quite distinct, but the Suppliers wish to make it clear that they will not go out themselves (or let any other person) and collect stocks of scarce or endangered butterflies or in any other way endanger our native fauna. Similarly livestock of endangered species will not be collected.

The formation of an Entomological Suppliers Association was proposed which will be co-operating in all such matters relating to Conservation Responsibility. At the meeting was a representative of the Ministry of Agriculture with whom there is also co-operation relating to importation of potential pest species.

Reproduced by kind permission of the AES (R. Baxter *et al.* 1976)

Appendices

APPENDIX 5: Insect re-establishment - a code of conservation practice

Introduction
The use of re-introductions and re-establishment of animals and plants, as part of projects aimed at re-creating habitats and communities, is widely accepted as constructive for the conservation of the countryside.

The Joint Committee for the Conservation of British Insects has been concerned at the lack of coordination, documentation or advice on appropriate techniques for the re-establishment of insects. Accordingly, it has produced this code of conduct, which it hopes will have wide application. It has consulted with other conservation organisations and is currently pressing the Nature Conservancy Council to produce a nationally accepted policy with guide lines for re-establishment and re-introduction.

This code of conduct has been agreed by the members of the Committee, representing the Royal Entomological Society, the Butterfly Conservation, The British Entomological and Natural History Society, the Amateur Entomologist's Society, the British Museum (Natural History), the IUCN(SSC) Butterfly specialist Group, and by observers of the English Nature, National Trust, Forestry Commission, Agricultural Development and Advisory Service and the Ministry of Defence on the Joint Committee.

1. Cautionary Foreword
Entomologists and conservationists are by no means agreed about the role re-establishment of invertebrates (see 'Definitions', 2. below) should play in the conservation of species and sites. Indeed, some insect conservationists believe that establishment of species may do more harm than good. Others are convinced that, under due safeguard, establishment of species has an increasingly important role in conservation. It is for these that this code is written. The Committee recommends that no specific proposal for insect re-establishment be condemned or approved without full discussion and consideration.

Any proposal to establish a population of insects must consider the objectives of doing so, together with the points for and against, including theoretical and practical ones. These cannot be set out fully in the code of practice, but the Committee is always willing to advise on particular cases.

However, the Committee believes that some ecological principles have been misunderstood in relation to establishment, and it urges that a thorough ecological assessment be made when considering the points for and against any establishment.

Appendix 5 cont..

Appendices

2. Definitions

Re-establishment means a deliberate release and encouragement of a species in an area where it formerly occurred but is now extinct. It is recommended that no species should be regarded as locally extinct unless it has not been seen there for at least five years.

Introduction means an attempt to establish a species in an area where it is not known to occur, or to have occurred.

Re-introduction means an attempt to establish a species in an area to which it has been introduced but where the introduction has been unsuccessful.

Reinforcement means an attempt to increase population size by releasing additional individuals to the population.

Translocation means the transfer of individuals from an endangered site to a protected or neutral one. Translocation is of less importance to insects than to longer-lived animals, such as mammals.

Establishment is a neutral term used to denote any attempt made artificially and intentionally to increase numbers of any insect species by the transfer of individuals.

3. Objectives

Objectives in establishing insect populations are many and varied. The three most important objectives are pest control, scientific research and wildlife conservation.

Biological, natural and integrated control are three types of pest management aimed at the establishment of insect populations. Biological control uses introductions, specifically. Establishments for pest control are not considered further in this code, though it may be helpful in planning them. Attention is drawn to the provisions of the Wildlife and Countryside Act 1981, which prohibit the introduction of alien species to the United Kingdom (Part 1, Section 14).

Establishments of insect populations for scientific research are often temporary, being made to elucidate some principle of scientific theory or practice. In most of its provisions, this code is relevant to this type of establishment.

Establishments of insect populations for conservation are arguably acceptable in principle, but are affected by individual circumstances, by the aims of conservation, and by considerations of geographical scale. Establishments cannot replace biotope conservation, or ensure conservation of species over their natural range.

Appendix 5 cont..

Appendices

Establishment of insect populations for conservation should focus particularly on the re-establishment of nationally threatened species, but the establishment of a particular resource, such as an attractive butterfly, for the enhancement of human enjoyment can also be considered. Re-establishments are particularly important because of recent trends in land-use (see 4. below)

It is recommended that for any proposed re-establishment, its objectives are clearly formulated, in detail, and made freely available for examination by responsible organisations (e.g. NCC, this Committee, BRC, DBCS). The need for confidentiality in particularly sensitive cases is recognised.

4. Trends in wildlife conservation

Whilst it is not the purpose of this code to advocate the use of re-establishments for conservation, the trend over the last 30 years has shown that they must be increasingly considered.

In the past, wildlife in some areas has been able to survive only because agriculture and forestry have been relatively inefficient in maximising yields of crops and timber.

Intensification of agriculture (and, to a lesser degree, forestry) has destroyed wildlife habitats over a wide area, leaving nature reserves as the most important wildlife refugia.

Nature reserves are a series of isolated and fragmented areas. Virtually all need to be managed to preserve their wildlife interest, but some have lost species through the lack of appropriate management. Some species may be particularly vulnerable to extinction in small reserves.

Although local extinctions and recolonisations have been the usual pattern in nature, the isolation of nature reserves makes recolonisation uncertain and unreliable.

The rehabilitation of nature reserves, and their creation from disused or abandoned land, may suggest the intervention of Man to establish wildlife in them.

Contrary to a widely held belief, many successful re-establishments have been made over the last few decades.

5. Planning for re-establishment

Re-establishment for conservation mat be species-orientated or site-orientated.

Appendix 5 cont..

Appendices

Species-orientated re-establishments are primarily aimed at endangered or vulnerable species whose very existence in the country is threatened by habitat destruction and change. Such species obviously merit particular attention. In some instances, it is appropriate also to consider introduction, in which case the risk of displacing other organisms should be considered.

Site-orientated re-establishments are usually aimed at enhancing the wildlife of a site (usually a nature reserve) by providing a showy, or otherwise valuable, species that was formerly present but has become extinct.

In practice, both site-orientated and species orientated re-establishments are dependent on adequate preparation of the site, or sites, to receive the species selected.

There is little point in attempting to re-establish species if its ecological requirements are not known or understood. It is recommended that every proposal for re-establishment states the detailed ecological needs of the species concerned and how they are to be met.

Although local extinctions may occur from a variety of events, a very common cause is simply lack of, or inappropriate, habitat management. Virtually no reserve (or other site) consists of 'climax' vegetation, and most are changing with time in the absence of management. It is recommended that no re-establishment be attempted unless the cause of extinction is well understood, and can be reversed. This is the counterpart to the paragraph above.

Before proceeding to prepare a site for re-establishment, it must be considered whether objections, theoretical and practical, have been given due weight. Is the proposed receiving site large enough? Will the re-established colony require constant reinforcement? Have genetic implications been fully thought out?

In the planning stage, an assessment of the impact of the proposed re-establishment on the receiving site should be prepared. Possible effects on other wildlife, especially species of conservation value, should be considered.

6. Preparing the receiving site

Permission to re-establish any species must be obtained from the owner-occupier of the designated site.

The adequacy of resources for the species on the receiving site should be determined, preferably through research.

Appendix 5 cont..

Appendices

The ecological conditions necessary for the re-established species must be imposed on the site before the re-establishment is attempted. Where continuous, regular or periodic management is required, this must be to an agreed, detailed plan, and the body attempting the re-establishment must be satisfied that management will proceed in accordance with the plan.

Re-establishment of any species, and the re-creation of its habitat, must be compatible with the objectives of management for the receiving site, and conform to the provisions of the management plan. Apparently incompatible objectives can often be achieved by suitable rotational management.

It is recommended that the attempted re-establishment be discussed fully with the site owner/occupier, and with the full reserve committee and scientific committee, as well as the warden in the case of nature reserves.

It is important to consult NCC because an SSSI may be involved. There are implications under the Wildlife and Countryside Act 1981, if this is the case.

7. The source of stock for re-establishment

An attempt at re-establishment must not weaken or harm the source population from which the stock is obtained. (Most colonies of insects, with a high rate of intrinsic natural increase, are able to withstand removal of stock, if their habitat is in a satisfactory condition).

Permission to take stock for re-establishment elsewhere must be obtained from the owner/occupier of the source site. The provisions of the Wildlife and Countryside Act, 1981, must be complied with. Advice can be obtained from regional officers of the Nature Conservancy Council.

The community of which the species for re-establishment is a part must be considered, and reproduced as far as possible on the receiving site. Specific parasites should be introduced with the source stock, if possible, as these are inevitably rarer, and therefore in even greater need of conservation than their hosts. An exception should of course be made where the purpose of the establishment is biological control rather than species conservation.

Stock of an ecological type most similar to that formerly inhabiting the receiving site should be chosen. Usually this will mean a source close to the receiving site, but not to the exclusion of other factors. Stock from a similar biotope should be preferred to a geographically closer but dissimilar biotope.

Appendix 5 cont..

Appendices

Consideration should be given to breeding in captivity stock for later release. In this way, numbers may be increased with less damage to the source.

The stage (egg, larva, pupa, imago) for release depends on circumstances; there is no generally applicable rule. Species with sedentary adults may be released with the exception that eggs will be laid in the most appropriate sites. Active adult insects may leave the site before oviposition. Larger numbers of immature stages than adults should be used in re-establishments, to allow for mortality between release and reproduction.

Numbers of released insects must be adequate to achieve re-establishment. Small numbers are often ineffective.

Detailed records of the exact procedures used in the attempt at re-establishment should be kept.

8. Monitoring re-establishments

All attempts at re-establishment, whether successful or not, should be reported to the Biological Records Centre (ITE, Monks Wood), and to this Committee. Confidentiality, if required, is assured. Secretive attempts can confuse others and result in lost information.

A standard form for recording re-establishments has been produced by this Committee, is available gratis from the Biological Records Centre, and should be sent, when completed, to the Committee's Surveys Officer. The relevant addresses are at 10 below.

Detailed assessment of the success of any attempt at re-establishment should be made, with continual re-assessment at frequent intervals. Such assessments should consider resources and other species.

In the case of butterfly re-establishments, success can be monitored using transect 'walks', undertaken during the adult flying period and compared with regional and national trends derived from the Butterfly Monitoring Scheme. Details may be obtained from this Committee or the organiser of the scheme, Dr. E. Pollard (ITE, Monks Wood Experimental Station).

As far as possible, re-establishments should be written up and published, so contributing to a common store of expertise.

Appendix 5 cont..

Appendices

9. Summary of main recommendations
1. Consult widely before deciding to attempt any re-establishment.
2. Every re-establishment should have a clear objective.
3. The ecology of the species to be re-established should be known.
4. Permission should be obtained to use both the receiving site and the source of material for re-establishment.
5. The receiving site should be appropriately managed.
6. Specific parasites should be included in re-establishment.
7. The numbers of insects released should be large enough to secure re-establishment.
8. Details of the release should be meticulously recorded.
9. The success of re-establishment should be continually assessed and adequately recorded.
10. All re-establishment should be reported to the Biological Records Centre and this Committee.

10. Useful addresses
Biological Records Centre, Monks Wood Experimental Station, Huntingdon, PE17 2LS.
Butterfly Monitoring Scheme, address as above.
JCCBI, c/o Royal Entomological Society of London, 41 Queen's Gate, London, SW7 5HU
English Nature, Northminster House, Peterborough, PE1 1UA
Butterfly Conservation, Tudor House, Quorn, Loughborough, Leicestershire, LE12 8AD
Amateur Entomologist's Society (Conservation Committee), 54 Cherry Way, Alton, Hampshire, GU34 2AX

Reproduced with kind permission of the Joint Committee for the Conservation of British Invertebrates.

Appendices

APPENDIX 6: Butterfly Conservation's Code of practice for butterfly releases

1. All establishment attempts conducted in the Society's name must conform with the Butterfly Conservation (BC) Policy on Butterfly Establishments and should be planned in liaison with the BC Conservation Sub-Committee.

2. All releases of butterflies, or of their immature stages, must aim to serve a clear nature conservation purpose. Releases should be both scientifically desirable and necessary. Butterflies should not be treated as amenity objects and releases should not be made purely for publicity purposes.

3. All releases must be properly planned.

4. Adequate knowledge of the status and distribution of butterflies in the region is essential before any establishment can be carried out.

5. Spur of the moment releases (e.g. of surplus breeding stock) must be resisted. Such poorly planned releases are likely to generate adverse comment on an issue which has a high emotive content and hence prove to be bad publicity for butterfly conservation as a whole.

6. It is essential to ensure that the habitat at the planned receiving site is fully suitable and can be maintained as such. Expert advice should be sought here. Detailed management plans are essential.

7. The ecological requirements of the species concerned must be adequately known before any establishment attempt can take place. Releases of poorly known species (such as the Large Tortoiseshell) should be restricted to bona fide scientific experiments carried out under supervision from English Nature or ITE.

8. Permission must be obtained from the landowner (or his agent) of both receiving site and the donor site.

9. Permission must be obtained from the local English Nature office if either the receiving or the donor site is an SSSI. If in doubt ask. Additionally, it is strongly advisable to consult NCC at regional level and the relevant County Naturalists' Trust about all possible establishment attempts.

Appendix 6 cont..

Appendices

10. Only native stock of indigenous species can be released into the wild. Releases of non-native species, sub-species or even foreign stock will contravene the 1981 Wildlife and Countryside Act.

11. Efforts must be made to ensure the release of a viable number of individuals at the most appropriate life-stage. The formula varies from species to species. Advice is available from the BC Conservation Sub-Committee.

12. Always use wild stock from a local source if at all possible. Ideally, from the nearest sizeable colony. Take particular care not to weaken the parent population, and avoid the transfer of individuals during seasons of scarcity.

13. Be careful not to ignore the above procedure when attempting to rescue a population from a site threatened by imminent development or destruction. If no suitable site for immediate translocation exists it is best to maintain a captive breeding nucleus until appropriate habitat can be created.

14. All releases must be reported to the Insect Establishment Recording Scheme, using the Scheme's standard form. Completed forms should be sent to the BRC at Monks Wood or to JCCBI at Wareham.

15. All establishment attempts must be closely monitored by an appropriate objective method. In some cases this will involve monitoring an immature stage. Much useful knowledge can be gained from clear information on establishment attempts, even from unsuccessful efforts.

Reproduced by kind permission of Butterfly Conservation (1994).

Appendices

APPENDIX 7: Butterfly Conservation's National Conservation Strategy - Summary

1. Butterfly Conservation's overriding objective is to ensure a future for our butterflies, moths and their habitats. We have grown to be the largest butterfly & moth conservation society in the world and have already received substantial donations to endow our future work. We will be working hard to continue the increase in membership and financial support. Some of our resources will be invested in achieving future growth, but we are determined that most of our annual expenditure will go on direct conservation work.

2. Successful butterfly conservation depends on detailed scientific knowledge. Butterfly Conservation is ideally placed to encourage its active members to take part in surveys aimed at gathering this knowledge and to increase the amount of professional scientific work that is devoted to butterfly and moth studies.

3. We know that nature reserves can play a vital part in protecting colonies or rare butterflies - but only if the reserve management is correct. As well as buying and managing more nature reserves of our own, Butterfly Conservation will be working closely with other nature reserve owners to encourage habitat management for butterflies and moths based on sound ecological principles. We hope many of our members will be able to play a direct role in this habitat management work.

4. Working nationally, we plan to identify the most threatened butterflies, moths and habitats as the basis of a national action plan. Different problems however affect the various regions of Britain. Butterfly Conservation will therefore develop regional action plans, identifying the most threatened species and sites in each part of Britain, and then strive to ensure that they are safeguarded. Butterfly Conservation's strong network of area groups will have a vital role in this work.

5. Unregulated and unplanned releases of butterflies do no good and may harm wild populations. On the other hand we know that properly-researched and planned introductions can re-populate a suitable habitat with species lost to the region. We are currently, among other projects, examining the possibility of restoring the Chequered Skipper to the English list of breeding butterflies.

Appendix 7 cont..

Appendices

6. Butterflies and moths are only effectively protected if their habitats are safeguarded. We will be campaigning for better legal protection of important habitats and the species that they contain. We will be bringing the needs of butterflies and moths to the attention of all those bodies which influence the way the countryside is managed, and promoting clear codes of behaviour for butterfly enthusiasts to follow when photographing or breeding the insects that they love.

7. Places rich in butterflies are normally rich in all forms of wildlife. We will be working closely with kindred conservation organisations to identify and protect key habitats and in Europe general. We see butterfly protection as part of a broader move towards more sustainable forms of agriculture and countryside management generally.

8. Everyone loves butterflies: increasing public awareness will help people understand how butterflies and moths fit into the network of nature. Butterfly Conservation will be continuing its educational work, with especial emphasis on encouraging an interest in butterflies amongst children.

This summary is dated March 1992, and was prepared by David Corke and approved by the Conservation Committee of Butterfly Conservation, with the aim of developing an effective wildlife conservation strategy. Reproduced with permission from Butterfly Conservation, 1994.

Appendices

APPENDIX 8: Nature Reserves for butterflies

Whilst some County Trusts did not respond to a request for information in the compilation of this Appendix, others had to be very selective. For instance, many of the 78 nature reserves owned or run by the Gloucester Wildlife Trust are rich in butterflies, but the following listed below are only their selected top six. Comments regarding the butterfly highlights of the nature reserves listed below have been deliberately omitted so as not to draw unnecessary attention to particularly sensitive species and habitats.

COUNTY

BERKSHIRE, BUCKINGHAMSHIRE AND OXFORDSHIRE NATURALISTS' TRUST (BBONT) see under separate counties

BERKSHIRE
Avery's Pightle
Blackwater Reach Meadow
Broadmoor Bottom
Hurley Chalk Pit
Seven Barrows
Watts Reserve

BRISTOL, BATH & AVON
Dolebury Warren
Folly Farm
Goblin Combe
Kingsdown Down
Walton Common

BUCKINGHAMSHIRE
Aston Clinton Ragpits
Butler's Hangings
Finemere Wood
Gomm Valley
Grangelands
Windsor Hill

Appendix 8 cont..

Appendices

CLWYD
Aberduna
Marford Quarry
Y Graig
Loggerheads Country Park
Graig Fawr
Cefn Yr Ogof
Fenns Moss NNR

CARMARTHENSHIRE
Pembrey Burrows LNR
Pembrey Country Park and Pembrey FNR
Dinefwr Castle Woods
Pendine Coast

CEREDIGION
Cwm Soden Nanternis
Llanafan Forestry Tracks
Rhos Pil-Bach Nature Reserve
Ynyslas Dunes
Ynyshir Nature Reserve

CUMBRIA
Latterbarrow
Whitbarrow, Hervey reserve,SSSI

CORNWALL
Bodmin Moor
Breney Common and Redmoor
Chapel Porth to St Agnes Beacon
Goss and Tregoss Moors NNR
Greenscombe Wood SSSI
Gwithian Dunes
Kynance Cove SSSI
Loggans Moor
Millook Woods
Penhale Dunes
Ventongimps
Welcombe & Marsland

Appendix 8 cont..

Appendices

DORSET
Fontmell Down SSSI
Hod & Hambledon Hills NNR,SSSI
Kingcombe Meadows
Powerstock Common SSSI
Piddles Wood

DURHAM WILDLIFE TRUST
Bishop Middleham Quarry
Low Barns, Witton-le-Wear,SSSI
Durham Coast SSSI

ESSEX
Abberton Reservoir
Fringringhoe Centre
Hanningfield Meadows
Horndon Meadows
Roding Valley Meadows
Waterhall Meadows

GLOUCESTERSHIRE
Betty Daw's Wood
Daneway Banks SSSI
Elliot / Swift's Hill SSSI
Foxes Bridge Bog
Siccaridge Wood
Woorgreens Lake and Marsh

GWYNEDD / NORTH WALES WILDLIFE TRUST
Great Orme County Park
Gogarth
Bryn Pydew
Moors around Bala
Newborough Warren NNR
Coedydd Aber NNR

KENT (East)
Park Gate Down
Lydden LNR
Orleston Forest

Appendix 8 cont..

Appendices

East Blean Wood
South Swale LNR
Sandwich Bay/Stoneless

KENT (West)
Burham Down
Darland Banks
Farningham Woods
Kemsing Downs
Queendown Warren LNR

ISLE OF MAN
Close Sartfield
The Chasms
Marine Drive

MONTGOMERYSHIRE WILDLIFE TRUST AND SHROPSHIRE WILDLIFE TRUST
Llanymynech Rocks SSSI

NORTHAMPTONSHIRE
Collyweston Quarries
Farthinghoe LNR
Glapthorn Cow Pastures
High Wood and Meadow
Salcey Forest

OXFORDSHIRE
Asham Meads
Dry Sandford Pit
Foxholes reserve
Glyme Valley Springs Hartslock
Hook Norton Railway Cutting Horley
Otmoor Rifle Range & Spinney
Sydlings Copse

PEMBROKESHIRE
Pengelli FNR
Welsh Wildlife Centre Nature Reserve
Westfield Pill Nature Reserve

Appendix 8 cont..

Appendices

SHROPSHIRE
Llanymynech Rocks & Llynclys Hill
Severn Valley Country Park
Whixall Moss NNR
Wyre Forest NNR

STAFFORDSHIRE
Kingswinford Railway, S.Staffs
Stafford-Newport railway

SUFFOLK WILDLIFE TRUST
Blaxhall Heath
Bonny Wood
Bradfield Woods
Lackford Pits
Landguard
Market Weston Fen
Wortham Ling

WILTSHIRE
Blackmoor Copse Nature Trail
Morgan's Hill Nature Trail
Middleston Down Reserve
Ravensroost Wood Reserve
Somerford Common Nature Trail
Whitesheet Hill Nature Trail

WORCESTERSHIRE
Brotheridge Green
Knapp and Papermill Reserve
Monkwood Reserve
Tiddesley Wood Reserve
Windmill Hill

APPENDIX 9 The British Species, by family

SKIPPERS
Chequered Skipper — Carterocephalus palaemon
Small Skipper — Thymelicus sylvestris
Essex Skipper — Thymelicus lineola
Lulworth Skipper — Thymelicus acteon
Silver-spotted Skipper — Hesperia comma
Large Skipper — Ochlodes venata
Dingy Skiper — Erynnis tages
Grizzled skipper — Pyrgus malvae

SWALLOWTAIL
Swallowtail — Papilio machaon

WHITES AND SUPHURS
Bath White — Pontia daplidice
Wood White — Leptidea sinapis
Clouded Yellow — Colias croceus
Berger's Clouded Yellow — Colias australis
Pale Clouded Yellow — Colias hyale
Brimstone — Gonepteryx rhamni
Large / Cabbage White — Pieris brassicae
Small White — Pieris rapae
Green-veined White — Pieris napi
Orange Tip — Anthocharis cardamines
Black-veined White — Aporia crataegi (extinct 1922)

BLUES, COPPERS, HAIRSTREAKS
Green Hairstreak — Callophrys rubi
Brown Hairstreak — Thecla betulae
Purple Hairstreak — Quercusia quercus
White-letter Hairstreak — Strymonida w-album
Black Hairstreak — Strymonidia pruni
Small Copper — Lycaena phlaeas
Large Copper — Lycaena dispar (extinct 1851)
Large Blue — Maculinea arion (extinct 1979)
Small/Little Blue — Cupido minimus
Silver-studded Blue — Plebejus argus
Brown Argus — Aricia agestis
Northern Brown Argus — Aricia artaxerxes
Common Blue — Polyommatus icarus
Chalkhill Blue — Lysandra coridon
Adonis Blue — Lysandra bellargus
Holly Blue — Celastrina argiolus
Short-tailed Blue — Everes argiades
Mazarine Blue — Cyaniris semiargus (extinct 1877)

DUKE OF BURGUNDY FRITILLARY
Duke of Burgundy Fritillary — Hamaeris lucina

Appendix 9 cont..

Appendices

ADMIRALS AND FRITILLARIES
White Admiral	Ladoga camilla
Purple Emperor	Apatura iris
Red Admiral	Vanessa atalanta
Painted Lady	Cynthia cardui
Camberwell Beauty	Nymphalis antiopa
Small Tortoiseshell	Aglais urticae
Large Tortoiseshell	Nymphalis polychloros
Peacock	Inachis io
Comma	Polygonia c-album
Small pearl-bordered Fritillary	Boloria selene
Pearl bordered Fritillary	Boloria euphrosyne
High Brown Fritillary	Argynnis adippe
Dark Green Fritillary	Argynnis aglaja
Silver-washed Fritillary	Argynnis paphia
Marsh Fritillary	Eurodryas aurinia
Glanville Fritillary	Melitaea cinxia
Heath Fritillary	Mellicta athalia
Queen of Spain Fritillary	Argynnis lathonia

MONARCH
Monarch or Milkweed	Danaus plexippus

BROWNS
Speckled Wood	Pararge aegeria
Wall Brown	Lasiomatta megera
Mountain Ringlet	Erebia epiphron
Scotch Argus	Erebia aethiops
Marbled White	Melanargia galathea
Grayling	Hipparchia semele
Gatekeeper or Hedge Brown	Pyronia tithonus
Meadow Brown	Maniola jurtina
Small Heath	Coenonympha pamphilus
Large Heath	Coenonympha tullia
Ringlet	Aphantopus hyperantus

ACRONYMS

AES	Amateur Entomologists' Society
BBCS	British Butterfly Conservation Society, became Butterfly Conservation
BC	Butterfly Conservation, formerly BBCS:British Butterfly Conservation Society
BCCPN	British Correlating Committee for the Protection of Nature
BMS	Butterfly Monitoring Scheme
BRC	Biological Records Centre
BUTT	Butterflies under Threat Team
CITES	Convention on International Trade in Endangered Species
CCCPN	Central Correlating Committee for the Protection of Nature
CCRESL	Conservation Committee of the Royal Entomological Society of London
CPBL	Committee for the Protection of British Lepidoptera
CPRE	Council for the Protection of Rural England
CRWS	Commons and Rights of Way Society
CTNC	Cornwall Trust for Nature Conservation
DNA	Deoxyribonucleic acid
DOE	Department of the Environment
EC	European Community, formerly EEC: European Economic Community
EN	English Nature, formerly NCC: Nature Conservancy Council
ES	Entomological Society of London
ESA	Environmentally Sensitive Area
EU	European Union, formerly European Community
FC	Forestry Commission
FCS	Favourable Conservation Status
FER	Forest Enterprise Reserve
FNR	Forest Nature Reserve
FOE	Friends of the Earth
GAP	General Agricultural Policy
GATT	General Agreement on Tariffs and Trade
ITE	Institute of Terrestrial Ecology
IUCN	International Union for the Conservation of Nature
JCCBI	Joint Committee for the Conservation of British Invertebrates (formerly Insects)
JNCC	Joint Nature Conservation Committee
JCCLBB	Joint Committee for the Conservation of the Large Blue Butterfly
LNR	Local Nature Reserve
MOD	Ministry of Defence

Acronyms cont..

NERC	Natural Environment Research Council
NHM	The Natural History Museum (London)
NC	Nature Conservancy (became NCC)
NCC	Nature Conservancy Council (became EN)
NGO	Non governmental Organization
NNR	National Nature Reserve
NRA	National Registered Authority
NRIC	Nature Reserves Investigation Committee
RDB	Red Data Book
RESL	Royal Entomological Society of London
RSNC	Royal Society for Nature Conservation, now "The Wildlife Trusts"
RSPB	Royal Society for the Protection of Birds
SCI	Site of Community Importance
SCPBL	Standing Committee for the Protection of British Insects
SPA	Special Protection Areas
SAC	Special Areas for Conservation
SNH	Scottish National Heritage
SPA	Special Protection Area
SPNR	Society for the Promotion of Nature Reserves (became RSNC)
SRP	Species Recovery Programme
ESA	Environmentally Sensitive Area
SSSI	Site of Special Scientific Interest
WWF	World Wide Fund for Nature

GLOSSARY

ACID RAIN DEPOSITION	Rain that includes sulphuric and nitric acid
ACIDIFICATION	Increase in acidity of soil or water
ALIEN, EXOTIC	An introduced species
AUTECOLOGY	The study of an organism in relation to its environment
BIODIVERSITY	The complement of all flora and fauna in an area
BIVOLTINE	Having two generations
CARR	A wetland woodland type of habitat, mostly of alder
CARRYING CAPACITY	The total number of individuals that any habitat can support
CLIMAX	A plant community that can perpetuate itself in the prevailing conditions of climate and soil type
CLINE	A continuous gradation of forms
DIRECT EFFECTS	A legal term indicating laws have direct action on targets
DIVERSITY	The ratio of species to numbers of individuals in a locality
ECOSYSTEM	A collection of habitats within a defined area
ENDANGERED	Species in danger of extinction and whose survival is unlikely if the causal factors continue operating
EUTROPHICATION	Enrichment of water by nutrients
EXTANT	Living
EXTENSIFICATION	Farming the land less
EXTINCT	Not having been seen alive in its locality for 50 years
GENE	Unit of heritable material
GENE POOL	The total number of genes represented by all individuals in a population
HABITAT	A place determined by climate, soil and organisms
HETEROZYGOUS	Having different gene pairs
HOMOZYGOUS	Having identical genes
IMMIGRATION	The one-way movement of animals, from A to B
IMPROVEMENT	Agricultural term for making land more productive
INDIRECT EFFECTS	A legal term indicating that laws do not have direct effects
INTENSIFICATION	Increased productivity from the land
INTRODUCTION	The deliberate or accidental release of an animal or plant species or race to a place where it has not occurred in historical times.
METAPOPULATION	A group of inter-linked populations

Glossary cont..

MIGRATION	The two-way movement of animals, from A to B to A
NATURALIZATION	The establishment of self-generated populations of an introduced species or race in a free-living in the wild.
NICHE	The role of a species in a habitat (not a place)
PHENOTYPE	The sum of the characteristics manifest by an organism
PLANT SUCCESSION	The natural series of events in which different sorts of vegetation succeed each other.
PRESERVATION	Looking after an individual species
RARE	Species with small world populations that are not at present 'endangered' or 'vulnerable', but are at risk; usually have localized populations
RE-INTRODUCTION	The deliberate or accidental release of a species or race into an area in which it was indigenous in historical times
RESPONSIBLE BODY	Usually a government body charged with authority
RESTOCKING	The deliberate or accidental release of a species or race into an area in which it it is already present.
RICHNESS	The number of species in a locality
SUPREMACY	The over-riding nature of some laws
TRANSLOCATION	An old term for moving a species from A to B
TRANSPLANTATION	An old term for moving a species from A to B.
UNIVOLTINE	Having a single generation, usually within a year.
VULNERABLE	Species believed likely to move into the 'endangered' category in the near future if the causal factors continue operating

Bibliography

ADAMS, W.M. & ROSE, C.J. (Eds) (1978) *The Selection of Reserves for Nature Conservation*. Discussion Papers in Conservation 20. University College London. 34pp.

AMATEUR ENTOMOLOGIST'S SOCIETY (1990) Habitat Conservation for Insects - A neglected Green Issue. *The Amateur Entomologist*, Volume 21: 1- 262pp. General Editor: Peter Cribb.

ANONYMOUS (1929) Report of the Committee appointed by the Entomological Society of London for the Protection of British Lepidoptera. *Proceedings of the Entomological Society of London*. 4: pp. 53-68.

ANONYMOUS (1986) Insect re-establishment - a code for conservation practice. *Antenna*. 10:13-18.

ANONYMOUS (1989) The impact on the terrestrial environment of constructing the Channel Tunnel. In. *The Natural Environment Research Council. Report of the Institute of Terrestrial Ecology for 1987/9.* pp. 3-4.

ARION (1993) A small extinction. *Butterfly Conservation News* (53) :6.

ASHER, J. (1994) *The Butterflies of Berkshire, Buckinghamshire and Oxfordshire:* Pisces Publications, Oxford. 160pp.

BAKER, P. (1994) The modified status of *Strymonidia w-album* (Knoch) (Lepidoptera:Lycaenidae) in North West Surrey. *British Journal of Entomology and Natural History*. February, 1994, Vol. 7, Part 1, 25-6.

BALFOUR, A.B. (1930) Butterflies and moths found in East Lothian. *Transactions of the East Lothian Antiquities and Field Naturalists' Society*. 1: 169-184.

BALFOUR-BROWNE, F. (1958a) The Origin of Our British Swallow-Tail and Our Large Copper Butterflies. *The Entomologist's Record & Journal of Variation*. 70: (1958) 33-36.

BALFOUR-BROWNE, F. (1958b) Keeping a species alive, some comments on man's attempts to re-establish wildlife in places he had earlier made uninhabitable for it. *The Field*, 2 January 1958.

BARKER, A. & BARKER, L. (1993) Aspects of oviposition and conservation of the Brown Argus butterfly. Lecture presented at international symposium on the Ecology and Conservation of Butterflies organised by Butterfly Conservation. 10-12 September 1993 at Keele University.

BARNHAM, M., FROGGITT, G.T. & RATCLIFFE, L.V. (1993) Recent changes in butterfly distributions in the Harrogate district. *Naturalist*. 118: 46-53.

BAXTER, R., THE BUTTERFLY FARM, L. CHRISTIE, D.B. JANSON, IMAGO BUTTERFLIES, NATURE OF THE WORLD, WATKINS & DONCASTER,

Bibliography

WORLDWIDE BUTTERFLIES LTD. (1976) A Code for Dealers. *Amateur Entomologist's Society Bulletin.* 35:4-5.

BENHAM, B.R. (1973) The declines (and fall?) of the Large Blue butterfly. *Bulletin of the Amateur Entomologist's Society.* 32:88-94.

BERRY, R.J. (1977) *Inheritance and Natural Selection.* Collins, London, New Naturalist Series.

BERRY, R.J. (1992) The role of ecological genetics in biological conservation. In *Conservation of Biodiversity for Sustainable Development.* O.T. Sandlund, K. Hindar & A.H.D. Brown (eds.) pp.107-123. Scandinavian University Press.

BINK, F.A. (1970) A review of the introductions of *Thersamonia dispar* Haw. (Lep.,Lycaenidae) and the speciation problem. *Entomologische Berichten.* 30: 179-183.

BIRKETT, N.L. (1995) Is the Silver-spotted Blue extinct in north-west England? *Butterfly Conservation News* 59:24-25.

BISSET, K. & FARMER, A.M. (1993) *SSSIs in England at risk from acid rain.* English Nature Science Series. No. 15. English Nature, Peterborough.

BLAB, J., RUCKSTUHL, T, ESCLE, T. & HOLZBERGER, R. (1987) *Aktion Schmetterlinge.* Ravenburger 192pp. (reprinted in Dutch and French).

BOURN, N. A.D. & THOMAS, J.A. (1992) The ecology and conservation of the Brown Argus butterfly, *Aricia agestis* in Britain. *Biological Conservation.* 63:67-74.

BRASSLEY, P. (1990) *Nature in Devon. The Journal of the Devon Wildlife Trust.* Devon Wildlife Trust, Exeter.

BREE, W. (1852) A list of Butterflies occurring in the Neighbourhood of Polebrooke, in the County of Northampton, with some remarks. *The Zoologist.* X:3348-3352.

BRETHERTON, R.F. (1951a) Our lost butterflies and moths. *Entomologist's Gazette* 2: 211-240.

BRETHERTON, R.F. (1951b) The early history of the Swallowtail butterfly (*Papilio machaon* L.) in England. *Entomologist's Record and Journal of Variation.* 63:206-211.

BRITISH BUTTERFLY CONSERVATION SOCIETY, (1988) *Butterflies of the Southern chalk downlands*: British Butterfly Conservation Society, pp.1-24.

BRITISH BUTTERFLY CONSERVATION SOCIETY, (1990) *25 Species of British Butterfly are Endangered.* A leaflet produced by the British Butterfly Conservation Society.

BRITISH BUTTERFLY CONSERVATION SOCIETY (1994) *BBCS Code and Policy on Butterfly Releases.*

BROOKS, S.J. (1993) Guidelines for Invertebrate Site Surveys; Joint Committee for the Conservation of British Invertebrates. *British Wildlife.* 4:(5) June 1993 pp.283-286.

BUTCHER,W.G. (1992) La cartographie des Lepidopteres Rhopaloceres rares et la gestion des milieux naturels en Somerset, Grande-Bretagne. *Actes du seminaire tenu au Mans les 6 et 7 novembre 1992, Museum National d'Histoire Naturelle, Paris, 1993.* pp. 137- 141.

BUTTERFLY CONSERVATION, (1992a) *Butterfly Conservation Education Pack.*

Bibliography

44pp of loose-leaf information on butterflies; produced by Butterfly Conservation.

BUTTERFLY CONSERVATION, (1992b) Help Save Catfield fen for the Swallowtails. *Butterfly Conservation News.* (51) Summer 1992 p. 32.

BUTTERFLY CONSERVATION, (1992c) *Introducing Butterfly Conservation.* Butterfly Conservation, sponsored by British Petroleum. 13pp.

BUTTERFLY CONSERVATION, (1992d) *Nature Conservation Strategy* - Summary. 2pp.

BUTTERFLY CONSERVATION, (1992e) *Nature Conservation Strategy.* 8pp.

BUTTERFLY CONSERVATION, (1994a) *The Land Rover Woodlands Campaign Information Pack, New Life for Old Woods:* Butterfly Conservation, sponsored by Land Rover.

BUTTERFLY CONSERVATION, (1994b) Beautiful Butterflies Bulldozed. Butterfly Conservation Press Information of 8 November, 1994. 3pp.

BUTTERFLY CONSERVATION, (1995a) Butterfly Conservation to the Rescue. *Butterfly Conservation Press Information* of 21 February 1995. 3pp.

BUTERFLY CONSERVATION, (1995b) Butterfly Conservation Press Information.

BUTTERFLIES UNDER THREAT TEAM (BUTT) (1986) *The Management of chalk grassland for butterflies.* Focus on Nature conservation, No. 17. Nature Conservancy Council, Peterborough.

CADBURY, J. (1990) The status and management of butterflies on RSPB reserves. *Conservation Review.* 4:40-46.

CADBURY, J. (1990) The Status and Management of Butterflies on RSPB Reserves. *RSPB Conservation Review* (4) Section 7: 40-46.

CARTER, C.I. & ANDERSON, M.A. (1987) *Enhancement of Lowland Forest ridesides and roadsides to benefit wild plants and butterflies.* Forestry Commission, Research Information Note 126: Forestry Commission Research Division.

C.I.T.E.S, 1991. see Convention on International Trade in endangered species of wild flora and fauna, and EC Council Regulation 1982.

CLARKE, C.A. (1954) Breeding the Large Blue Butterfly in Captivity, 1953-54. *Entomologist's Record and Journal of Variation.* Vol. 66 p. 209-211.

CLARKE, H. (1991) *The butterflies of Jersey: a conservation report.* 83pp. Dissertation presented for Conservation Course, University College, London.

CLARKE, S.A. & ROBERTSON, P.A. (1993) The relative effects of woodland management and pheasant predation on the survival of pearl-bordered fritillaries (*B. euphrosyne* and *B. selene*) in the South of England. *Biological Conservation.* 65:199-203.

COLLIER, R.V. (1978) *The status and decline of butterflies on Castor Hanglands NNR 1919-1977.* Unpublished report to Nature Conservany Council.

COLLIER, R.V. (1986) *The Conservation of the Chequered Skipper in Britain.* Focus on Nature Conservation. 16, Nature Conservancy Council, Peterborough.

COLLINS, N.M. (1987a) *Butterfly houses in Britain, the conservation implications.*

Bibliography

IUCN, Cambridge, UK. 60pp.

COLLINS, N.M. (1987b) *Legislation to conserve insects in Europe*. Pamphlet No. 13. 80pp. Amateur Entomologist's Society

COLLINS, N.M & MORRIS, M.G. (1985) *Threatened swallowtail butterflies of the World, The IUCN Red Data Book.* IUCN, Switzerland, Gland & Cambridge, UK. 432pp.

COLLINS, N.M & THOMAS, J.A. (1991) *The Conservation of Insects and Their Habitats*. London, Academic Press. 450pp.

COLLINS, N.M., MORRIS, M.G. & WHALLEY, P. (1988) The evolution of concern: conservation and the Royal Entomological Society. *Antenna*, 12:158-163.

COMMAND REPORT No. 7122. *Report on the Conservation of Nature in England and Wales;* presented to Parliament July 1947.

COMMITTEE FOR THE PROTECTION OF BRITISH LEPIDOPTERA, Minutes of, from 1925 to 1931.

CONVENTION CONCERNING THE PROTECTION OF THE WORLD CULTURAL AND NATURAL HERITAGE, The "World Heritage Convention", 1972.

CONVENTION ON INTERNATIONAL TRADE IN ENDANGERED SPECIES (C.I.T.E.S.) (1991).

CONVENTION ON THE CONSERVATION OF MIGRATORY SPECIES OF WILD ANIMALS, (The "Bonn Convention") 1979.

CONVENTION ON WETLANDS OF INTERNATIONAL IMPORTANCE ESPECIALLY AS WATERFOWL HABITAT (1979), otherwise known as the Ramsar Convention.

CORKE, D., (1985) Is Essex a Poor County for Butterflies? p. 16-21 In. *The Larger Moths and Butterflies of Essex*. Edited by EMMET, A.M., PYMAN, G.A. & CORKE, D., Essex Naturalist (8) 135pp.

CRICHTON, M.& NI'LAMHNA, E. (1975) *Provisional Atlas of Butterflies in Ireland. (part of The European Invertebrate Survey)*. Irish Biological Records Centre, Dublin.

CROUCHER, J. P. (1992) The Status of Terrestrial and Freshwater Invertebrates. Population Monitoring in Britain and Ireland; A survey. English Nature Research Report No.24 (Report 94pp and Appendix A *xiii*) and Report No.25 (Appendix B:not paginated, but over 200pp).

DAVIDSON, J. & LLOYD, R. (1977) *Conservation and Agriculture*. John Wiley & Sons Ltd., Chichester, New York, Brisbane, Toronto.

DAVIES, M. (1992) *The White-letter Hairstreak Butterfly*. British Butterfly Conservation Society. 27pp.

DAVIS, B.N.K. (1989a) Habitat creation for butterflies on a landfill site. *Entomologist*. 108 1-2 pp. 109-122.

DAVIS, B.N.K. (1989b) Creation of buterfly habitats on a landfill site. In. *The Natural Environment Research Council, Report of the Council for the period 1 April 1985-31 March 1986., Report of the Institute of Terrestrial Ecology for 1985-1986*. Institute of

Bibliography

Terrestrial Ecology, pp.82-84.
DAVIS, B.N.K., LAKHANI, K.H. & YATES, T.J. (1991) The hazards of insecticides to butterflies of field margins. *Agriculture, Ecosystems and Environment.* pp.151-161.
DEMPSTER, J.P. (1977) Ecology of the swallowtail butterfly. In. *Annual Report 1976,* Institute of Terrestrial Ecology. pp.39-40.
DEMPSTER, J.P. (1993) The ecology and conservation of the Swallowtail butterfly (*Papilio machaon* L.) in Britain. Lecture presented at international symposium on the Ecology and Conservation of Butterflies organised by Butterfly Conservation. 10-12 September 1993 at Keele University, and (1995) In. *Ecology and Conservation of Butterflies*, Edited by A.S. Pullin. pp.137-149.
DEMPSTER, J.P. & HALL, L. (1980) An attempt at re-establishing the swallowtail butterfly at Wicken Fen. *Ecological Entomology.* (1980) 5:327-334.
DEMPSTER, J.P., KING, M.L. & LAKHANI, K.H. (1976) The status of the swallowtail butterfly in Britain. *Ecological Entomology.* (1976) 1: 71- 84.
DENNIS, R.L.H. (1977) *The British Butterflies, Their Origin and Establishment.* E.W.Classey Ltd., Faringdon, Oxon., 318pp.
DENNIS, R.L.H., (1992) *The Ecology of Butterflies in Britain.* Oxford University Press, Oxford. 354pp.
DENNIS, R.L.H. (1993) *Butterflies and Climate Change.* Manchester University Press, Manchester & New York. 302pp.
DEPARTMENT OF THE ENVIRONMENT, (1992) *Chronological list of EEC Environmental Directives, Decisions and Regulation. London, Department of the Environment* (unpublished and regularly updated document).
DOVER, J.W. (1991) The Conservation of Insects on Arable Farmland. pp.293-315. In. Collins, N.M. & Thomas, J.A. (eds.). *The conservation of insects and their habitats*, Academic Press, London.
DOWDESWELL, W.H. (1981) *The Life of the Meadow Brown.* Heinemann Educational Books, London. 165pp.
DUFFEY, E. (1968) Ecological studies on the Large Copper butterfly *Lycaena dispar* Haw. *batavus* Obth. at Woodwalton Fen National Nature Reserve, Huntingdonshire. *Journal of Appied Ecology.* 5: 69-96.
DUFFEY, E. (1970a) *Conservation of Nature.* Collins, London. 128pp.
DUFFEY, E. (1970b) Some effects of summer floods on Woodwalton Fen in 1968/69. *Entomologist's Gazette.* 21: (1) January 1970. 23 -26.
DUFFEY, E. (1974) *Nature Reserves and Wildlife.* Heinemann Educational Books, London. 134pp.
DUFFEY, E. (1977) Re-establishment of the Large Copper butterfly *Lycaena dispar batava* Obth. on Woodwalton Fen National Nature Reserve, Cambridge, England. *Biological Conservation.* 12: (2) 143-158.
DUFFEY, E. & MASON, G. (1970) Some effects of summer floods on Woodwalton Fen in 1968/69. *Entomologist's Gazette.* 21: p. 23-26.

Bibliography

DUFFEY, E. & WATTS, A.S. (eds.) (1971) The Scientific Management of Animal and Plant Communities for Conservation. *British Ecological Society Symposium* No. 11.Blackwell, Oxford.

DUNBAR, D. (1994) *Saving Butterflies, a practical guide to the Conservation of Butterflies.* Butterfly Conservation. 80pp.

EHRLICH, P.R. (1983) Genetics and the extinction of butterfly populations. In. *Genetics and Conservation. a reference for managing wild animals and plant populations.* C.M. Schonewald-Cox, S.M. Chambers, B. MacBryde and L. Thomas (eds.) Menlow Park, Benjamin/Cummins.

ELIAS, D.O.E., POLLARD, E, SKELTON, M.J. & THOMAS, J.A.T. (1974) A method of assessing the abundance of butterflies in Monks Wood National Nature Reserve in 1973. *Entomologists' Gazette.* 26:79-88.

ELLIS, E.A., 1965 *The Broads.* London, Collins..

ELMES, G.W, & THOMAS, J.S. (1992) Complexity of species conservation in managed habitats: Interaction between *Maculinea* butterflies and their ant hosts. *Biodiversity & Conservation.* 1 (3) p. 155-169.

ELTON, C.S. (1961) *Reserves for the Black Hairstreak butterfly in Bernwood Forest, Oxon/Bucks.* Unpublished. Nature Conservancy Council, Peterborough.

EMMET, A. M. & HEATH, J. (1989) *The Moths and Butterflies of Great Britain and Ireland. Vol. 7. Pt.1. Hesperiidae-Nymphalidae. The Butterflies.* Harley Books, Colchester. p.76-81.

ENGLISH NATURE, (1992) *1st Report, 1st April 1991 - 31st March 1992.* 124pp.

ENGLISH NATURE, (1993a) *The Large Blue is back in business: Species recovery programme.* English Nature. (6) March, 1993. p. 4-5.

ENGLISH NATURE, (1993b) *2nd Report, 1st April 1992 - 31st March 1993.* 115pp

ENGLISH NATURE, (1993c) *Facts and Figures, Information Guide,* leaflet. Autumn 1993.

ENGLISH NATURE, (1993d) *Large Blue Butterfly,* Series leaflet as part of Species Recovery Programme. (March, 1993).

ENGLISH NATURE, (1993e) *English Nature Science. No. 17. English Nature's research programme, 1992/3.* English Nature. 53pp.

ENGLISH NATURE (1994a) *National Nature Reserves / Marine Nature Reserves.* Report edited by Eddie Idle. English Nature, Peterborough.

ENGLISH NATURE (1994b) 3rd Report, 1st April 1993 to 31st March 1994. English Nature, Peterborough.

ENTOMOLOGICAL LIAISON COMMITTEE. (1964) Meeting of this committee, dated 10 November 1964.

ERHARDT, A. & THOMAS, J.A. (1991) Lepidoptera as indicators of change in the Semi-natural grasslands of Lowland and Upland Europe. pp. 213-237. In. Collins, N.M. & Thomas, J.A. (eds). *The conservation of insects and their habitats,* Academic Press, London.

Bibliography

EC Council Regulation 1982 (EEC) No. 3626/82 of 3 December 1982 on the implementation in the Community of the Convention on international trade in endangered species of wild fauna and flora.

EC Directive 79/409/EEC of 25 April, 1979 on the conservation of wild birds.

EC Directive 92/43/EEC of 21 May 1992 on the conservation of natural habitats and of wild fauna and flora (otherwise called the Habitats Directive).

EUROPEAN COMMUNITY, (1992) Eurostat, basic statistics of the Community 29th edition, 1992. Office for Official Publications of the European Community. 331pp.

EVERSHAM, B.C. (1994) Using invertebrates to monitor land use change and site management. *British Journal of Entomology and Natural History*, February 1994, Vol.7,(Supplement 1) p.36-45.

FARRELL, L. (1973) *A preliminary report on the status of the Chequered skipper (Carterocephalus palaemon (Pall.)) in the British Isles.* September 1973. Nature Conservancy Council, Peterborough.

FARRELL, L. (1976) Retreat to the North: the decline of the chequered skipper. *Country Life*. 19 August. 160:491.

FEARNEHOUGH, T. D. (1938) Hunting the Swallowtail. *The Journal of the Amateur Entomologist's Society*. 3: (27) p.52.

FEBER, R.E. (1993) *The ecology and conservation of buterflies on lowland arable farmland.* D. Phil. thesis. University of Oxford.

FEBER, R. & SMITH, H. (1993) Managing farmland for butterflies. Lecture presented at international symposium on the Ecology and Conservation of Butterflies organized by Butterfly Conservation. 10-12 September 1993 at Keele University, and (1995) as Butterfly conservation on arable farmland. In. *Ecology and Conservation of Butterflies*, Edited by A.S. Pullin. pp.84-97. .

FELTWELL, J. (1982) *Large White Butterfly. The Biology, Biochemistry and Physiology of Pieris brassicae (Linneaus)* Dr.W.Junk Publishers, The Hague, Boston, London. Volume 18. In. *Series Entomologica*. 535pp.

FELTWELL, J. (1983) Butterfly behaviour - *Celtis, Crataegi, Spini*. Notes and Observations. *Entomologist's Record and Journal of Variation*. 95:169-170.

FELTWELL, J. (1984) *Field Guide to the Butterflies and other insects of Britain.* Reader's Digest, London. 352pp.

FELTWELL, J. (1985) *Discovering Doorstep Wildlife.* Hamlyn, London. 160pp.

FELTWELL, J. (1986) *The Natural History of Butterflies.* Croom Helm, Orpington. 133pp.

FELTWELL, J. (1987) *The Naturalist's Garden.* Ebury Press, London. 160pp.

FELTWELL, J. (1989) *A Guide to Countryside Conservation.* Ward Lock, London. 159pp.

FELTWELL, J. (1990) *Butterflies, A practial guide to their study in school ground via the national curriculum.* Learning Through Landscapes Trust, Winchester. 20pp.

FELTWELL, J. (1992) *Meadows, A History and Natural History*. Alan Sutton, Stroud.

Bibliography

206pp.
FELTWELL, J. (1994) Beyond caterpillar tracks. *Land Rover World.* 9:78-81.
FELTWELL, J. (1995) Conserving butterflies in the Highlands and Dales. *Good Motoring.* April / June 36-37.
FELTWELL, J. & COPPIN, N. (1989) Unpublished report on the status of the main habitats, main issues, and endangered species in Cornwall. Sponsored by the World Wide Fund for Nature, the Cornwall Trust for Nature Conservation and English China Clays.
FELTWELL, J. & PHILP, E. (1980) Natural History of the M20 Motorway. *Transactions of the Kent Field Club.* 8(2)101-114.
FITTER, A. (1978) *An Atlas of the Wild Flowers of Britain and Northern Europe.* Collins, London. 272pp.
FITTER, R.S.R. (1994) The post-war conservation committees. *Bulletin of the British Ecological Society.* 25: (4) 205.
FORD, E.B. (1943) Two rare Cumberland butterflies. *Proceedings of the Royal Entomological Society of London.* (A) p. 18 parts 1-3 (May, 1943).
FORD, E.B. (1945) *Butterflies.* Collins, London. New Naturalist Series. 368pp.
FORD, E.B. (1948) Letter dated 15 February 1948 to Norman Riley in JCCBI archive at RESL.
FORD, E.B., (1965) *Ecological Genetics.* 2nd edition. Methuen, London.
FOWLES, A.P. (1991) Lepidoptera recording in Ceredigion, 1991. *Dyfed Invertebrate Group Newsletter.*
FRAZER, J.F.D. (1967) Insect Introductions. In. *The Biotic effects of Public Pressures on the Environment.* the Third Scientific Staff Symposium held at Monks Wood Experimental Station, March 20-21, 1967.
FRIDAY, L.R. (1993) *Wicken fen - the making of a Wetland Nature Reserve.* Harley Books, Colchester. 288pp.
FROHAWK, F.W.(1924) *Natural History of Butterflies.* (2 vols). Hutchinson, London.
FRYER, J.C.F.(1925) Protection of British butterflies. *Entomologist* 58:12-15
FULLER, R.J. & WARREN, M.S. (1990) *Coppiced woodlands - their management for wildlife.* Nature Conservancy Council, Peterborough. 34pp.
FULLER, R.J. & WARREN, M.S. (1991) Conservation management in ancient and modern woodlands: responses of fauna to edges and rotations. In. Spellerberg, I.F. Goldsmith, F.B. & Morris, M.G. (eds). *The Scientific Management of Temperate Communities for Conservation.* 445-472. Blackwell Scientific, Oxford.
GARDINER, B.O.C. (1960) Swallowtails for Wicken. *The Countryman.* 57:(2) 294-297.
GARDINER, B.O.C. (1963) Notes on the Breeding and Biology of *Papilio machaon* L. (Lepidoptera:Papilionidae). *Proceedings of the Royal Entomological Society of London.* (A)38:pts.10-12. 206-211.
GARDINER, B.O.C. (1971) Commercial Entomology - or is One Man's Rarity Another's Livelihood? *Entomologist's Record and Journal of Variation.* 89:5-34.

Bibliography

GARDINER, B.O.C. (1976) Collecting Controversey - facts wanted. *Entomologist's Record and Journal of Variation.* 88:110-112.

GARDINER, B.O.C. (1991) The swallowtail (*Papilio machaon*) and Large copper (*Lycaena dispar batavus*) at Wicken. *Entomologist's Record and Journal of Variation.* 103:200.

GARDINER, B.O.C. (1993) Father and son: The Newmans and their Kent Butterfly Farm. *Entomologist's Record and Journal of Variation.* 105:105-114.

GOODALL, A. (1983) Wildlife and Conservation in Lincolnshire. pp.51-58. In. *The Butterflies and Larger Moths of Lincolnshire and South Humberside.* Lincolnshire Natural History Brochure No. 10. Lincolnshire Naturalist's Union, Lincoln. Edited by Duddington, J., Johnson, R. 299pp.

GOODDEN, R. & GOODDEN, R. (1973) *The Large Blue. Report No. 1.* Worldwide Butterflies Ltd. 2pp.

GOODDEN, R. & GOODDEN, R. (1974) Saving the Large Blue, Britain's rarest butterfly. *Country Life.* 4 July 1974.

GOSS, H. (1884) On the probable early extinction of *Lycaena arion* in Britain. *Entomologist's Monthly Magazine.* 21:107-109.

GREATOREX-DAVIS, J.N. (1986) Management guidelines for the conservation of invertebrates, especially butterflies, in plantation woodland. (CSD report no.622) Nature Conservancy Council, Peterborough.

GREATOREX-DAVIS, J.N., SPARKS, T.H., HALL, M.L. and MARRS, R.H. (1993) The influence of shade on butterflies in rides of coniferised lowland woods in southern England and implications for conservation management. *Biological Conservation.* 63:31-41.

HABITATS DIRECTIVE, see EC Directive 92/43/EEC.

HABITAT (1979) Extinction of the Large Blue. 15 September, p. 3.

HALL, M. (1981a) *Butterfly research in I.T.E.* Cambridge. 28pp

HALL, M.L. (1981b) *Butterfly Monitoring Scheme, Instructions for independent recorders.* Institute of Terrrestrial Ecology (NERC), Cambridge.

HALL, M.L., GREATOREX-DAVIES, N. & POLLARD, E. (1989) Management of conifer plantations for invertebrates, especially butterflies. In. *The Natural Environment Research Council, Report of the Council for the period 1 April 1985-31 March 1986.* pp. 77-78

HAMBLER, C. & SPEIGHT, M.R. (1995) Biodiversity conservation in Britain: Scientific replacing tradition. *British Wildlife* 6: (3) 137-147.

HARDING, P.T. & GREEN, S.V. (1991) *Recent surveys and research on butterflies in Britain and Ireland: a species index and bibliography.* 42pp.

HARDING, P.T., ASHER, J. & YATES, P. (1993) Butterfly Monitoring 1 - recording the changes. Lecture presented at international symposium on the Ecology and Conservation of Butterflies organized by Butterfly Conservation. 10-12 September 1993 at Keele University.

Bibliography

HARRIS, MOSES, 1766. *The Aurelian.*
HEATH, J. (1970) *Provisional Atlas of the Insects of the British Isles. Part 1. Lepidoptera Rhopalocera, Butterflies. Maps 1-57.* Biological Records Centre, Huntingdon.
HEATH, J. (1975) *Insect Distribution Maps Scheme.* Lepidoptera Rhopalocera Butterflies. Provisional Distribution Maps. Biological Records Centre, Huntingdon.
HEATH, J. (1981) *Threatened Rhopalocera (Butterflies) in Europe.* European Committee for the Conservation of Nature and Natural Resources. Council of Europe.
HEATH, J. (1982) *Distribution Maps of the Butterflies of the British Isles.* Biological Records Centre, Huntingdon, Cambridge.
HEATH, J., POLLARD, E. & THOMAS, J. (1984) *Atlas of Butterflies in Britain and Ireland.* Viking, published in association with the Natural Environmental Research Council and Nature Conservancy Council, Peterborough. 158pp.
HEDRICK, P.W.& MILLER, P. (1992) Conservation genetics: theory and management of captive populations. In. *Conservation of Biodiversity for Sustainable Development.* edited by O.T. Sandlund, K. Hindar and A.H.D. Brown. Scandinavian University Press. pp.55-69.
HER MAJESTY'S STATIONERY OFFICE, (1975) *Conservation of Wild Creatures and Wild Plants Act 1975.* Her Majesty's Stationery Office, London.
HER MAJESTY'S STATIONERY OFFICE, (1981) *Wildlife and Countryside Act 1981,* Chapter 69. Her Majesty's Stationery Office, London. 128pp.
HER MAJESTY'S STATIONERY OFFICE, (1994) *Biodiversity, The UK Action Plan.* January 1994. Her Majesty's Stationery Office, London. 188pp.
HESLOP, I.R.P. (1967) A 1958 survey of the butterflies of Blackmoor Copse National Reserve. *Entomologist's Record and Journal of Variation.* 79:296-302.
HIGGINS, L., HARGREAVES, B. & LHONORE, J. (1991) *Guide complet des Papillons d'Europe et d'Afrique du Nord.* Delachaux et Niestlé, Paris. 270pp.
HOWARTH, T.G. (1973) The Conservation of the Large Blue Butterfly (*Maculinea arion* L.) in West Devon and Cornwall. *Proceedings of the British Entomological and Natural History Society.* 5: (4) 121-126.
HUNT, O.D. (1965) *Status and conservation of the Large Blue butterfly, Maculinea arion. L. 'The Conservation of Invertebrates',* Nature Conservancy Council, Monks Wood.
INSTITUTE OF TERRESTRIAL ECOLOGY (N.E.R.C.) *Annual Report*s from 1974 onwards. London, Her Majesty's Stationery Office.
INTERNATIONAL UNION FOR CONSERVATION OF NATURE AND NATURAL RESOURCES. (1986). *The IUCN position statement on translocation of living organisms, introductions, re-introductions and re-stocking.* Prepared jointly by the Species Survival Commission and the Commission on Ecology.
IRWIN, A.G. (1984) The large copper, *Lycaena dispar dispar* (Haworth) in the Norfolk Broads. *Entomologist's Gazette and Journal of Variation.* 96: 9-10, 212-213.

Bibliography

JACKMAN, R. (1981) Butterflies on the Brink. *Sunday Times*. 9 August supplement.

JOINT COMMITTEE FOR THE CONSERVATION OF BRITISH INSECTS (1972a) British Macrolepidoptera, Rare and endangered species and subspecies list. *Entomologists' Gazette*. 23: 278-279.

JOINT COMMITTEE FOR THE CONSERVATION OF BRITISH INSECTS. (1972b) *A code for insect collecting*. Joint Committee for the Conservation of British Insects.

JOINT COMMITTEE FOR THE CONSERVATION OF BRITISH INSECTS. (1972c) British Macrolepidoptera: rare and endangered species and forms. *Entomologist's Record & Journal of Variation*. 84: 210-212.

JOINT COMMITTEE FOR THE CONSERVATION OF BRITISH INSECTS. (1974) Rare and endangered species: general list. *Entomologist's Monthly Magazine*. 109: 200-201.

JOINT COMMITTEE FOR THE CONSERVATION OF BRITISH INSECTS (1987) Press Statement dated 15 July. *The return of the Large Blue butterfly*. 2pp.

JONES, M. (1986) The large blue is back where it belongs. *New Scientist*. 11 December. JOY, J. (1992) *Shropshire silver-studded blue, the Conservation of the Slver-studded blue (Plebejus argus) of Shropshire*. English Nature/World Wide Fund for Nature.

JOY, J. (1993) *The Ecology and Life History of the Large Heath butterfly in Shropshire*.

JOY, J. & WILLIAMS, S.M. (1990) Large Heath under threat. *Butterfly News* 46: Autumn/Winter 1990. p. 34-38.

JOYCE, C. (1988) Global warming could wipe out wildlife. *New Scientist*. 4 February p.29.

KIRBY, P. (1992) *Habitat Management for Invertebrates: a practical handbook*. Royal Society for the Protection of Birds, Sandy, Beds. 150pp.

KIRKLAND, P. (1995a) The Beleagured Marsh Fritillary. Butterfly Consrvation News. 59:6-7.

KIRKLAND, P. (1995b) A review of the distribution, ecology and behaviour of the scotch argus. *The Bulletin of the British Ecological Society*. 26:95-102.

KUDRNA, O. (1985) *Butterflies of Europe*. Vol. 1. Concise Bibliography of European Butterflies. AULA-Verlag GmbH, Wiesbaden. 447pp.

KUDRNA, O. (1986) *Butterflies of Europe*. Vol.8. Aspects of the Conservation of Butterflies in Europe. Edited by Otakar Kudrna. AULAG-Verlag, Wiesbaden. p. 323.

LABOUCHERE, F.A. (1935) Report on visit to Bude. (dated 31 July 1935) 4pp In: *Minutes of the Committee for the Protection of British Lepidoptera*.

LABOUCHERE, F.A. (1942) *Letter* to Edelsten dated 3 November 1942 In: JCCBI archives in RESL.

LAWSON, T. (1991) Save the butterfly and the botany. *The Independent* 1 April.

LACY, R.C. (1988) A report on population genetics in conservation. *Conservation Biology*. 2: 245-247.

LINNEAN SOCIETY, (1993) Biological Diversity, in *The Linnean, Newsletter and*

Bibliography

Proceedings of the Linnean Society of London. 9: (1) Jan, 1993, p. 13-14.

LORIMER, R.I. (1983) *The Lepidoptera of the Orkney Islands.* E.W.Classey Ltd., Middlesex. See section on Conservation p.9-10.

LUQUET, G-C. (1988) *Sauvons les Papillons, Les connaître pour mieux les protéger.* French Edition (translated from the German) by Duculot, Paris & Gembloux. 192pp.

LYSTER, S. (1985) *International Wildlife Law. An analysis of international treaties concerned with the conservation of wildlife.* Grotius Publications Limited. Cambridge.

McLEAN, I.F.C. (1990) The 1988 Presidential Address. Part 2. What Future for our Entomological heritage. *British Journal of entomology and Natural History.* 3: p. 35-54.

McLEAN, I.F.G. FOWLES, A.P., KERR, A.J., YOUNG, M.R. & YATES, T.J. (1993) Butterflies on nature reserves in Britain. Lecture presented at international symposium on the Ecology and Conservation of Butterflies organised by Butterfly Conservation. 10-12 September 1993 at Keele University, and (1995) In. *Ecology and Conservation of Butterflies*, Edited by A.S. Pullin. pp.68-83.

MAHON, A. & PEARMAN, D. (1993) *Endangered Wildlife in Dorset.* Dorset Environmental Records Centre. 134pp.

MARSDEN, H. (1884) On the probable extinction of arion in England. *Entomologists' monthly Magazine.* 21:186-189.

MAY, R. (1993) Biological diversity: Britain's traditions and their implications. In. *Action for biodiversity in the UK, A seminar organised by the Joint Nature Conservation Committee and the Department of the Environment.* edited by F.J. Wright, C.A. Galbraith and R. Bendall. JNCC. pp.16-20.

MEASURES, D.G. (1976) *Bright Wings of Summer.* London, Cassell.

MENDEL, H. & PIOTROWSKI, S.H. (1986) *The Butterflies of Suffolk, an atlas and history.* Suffolk Naturalist's Society. pp.128. see Chapter on Conservation. pp.51-58.

MILES, S.R. (1995) Report of the discussion meeting held on 12 May 1992 to consider invertebrate conservation in the United Kingdom. *British Entomological and Natural History Society.* 8:19-26.

MILLER, W.E. (1979) Fire as an insect management tool. *Bulletin of the Entomological Society of America.* 25: 137-140.

MONKS WOOD EXPERIMENTAL STATION. *Annual Reports.* Reports 1960-1965, 1966-1968, 1969-1971, 1972-1973 The Nature Conservancy, Monks Wood.

MOORBATH, P. (1973) The Conservation of Britain's Rarest Butterfly, *Maculinea arion* in South-East England. (Unpublished manuscript in English Nature archives, Peterborough).

MOORE, N.W. (1959) Welcombe Valley, Devon. Memorandum by the Regional Officer for the South West. internal document Nature Conservancy.

MOORE, N.W. (1987) *The bird of time, The science and politics of nature conservation.* Cambridge University Press, Cambridge. 290pp.

MORGAN, I.K. (1989) *A Provisional Review of The Butterflies of Carmarthenshire.* Nature Conservancy Council (unpublished). 65pp.

MORGAN, I.K. (1990-1993) Notes on lepidoptera, in *Dyfed Invertebrate Group Newsletter*.
MORIARTY, F. (1969) Butterflies and insecticides. *Entomologist's Record and Journal of Variation.* 81:276-278.
MORRIS, F.O. (ed.) (1895) *A History of British Butterflies*. John Nimmo, London.
MORRIS, M.G. (1967a) Insect collecting with special reference to nature reserves. In. Duffey, E. *The Biotic Effects of Public Pressures on the Environment.* Monks Wood Experimental Station, Abbots Ripton.
MORRIS, M.G. (1967b) The representation of butterflies (Lep., Rhopalocera) on British Statutory Nature Reserves. *Entomologist's Gazette.* 18: 57-68
MORRIS, M.G. (1968) Erroneous records of butterflies (Lep., Rhopalocera) from National Nature Reserves. *Entomologist's Gazette* 19:150.
MORRIS, M.G. (1976) Conservation and the collector. Chapter 6, pp. 107-116. In. *The Moths and Butterflies of Great Britain and Ireland.* Volume 1. Micropterigidae - Heliozelidae. Edited by John Heath. Blackwell Scientific Publications Ltd and The Curwen Press Ltd., London.
MORRIS, M.G. (1977) Entomology at Furzebrook Research Station, Dorset, England. *Antenna* July 1977 (1) 1: 6,8.
MORRIS, M.G. (1981) *Conservation of butterflies in the United Kingdom. Beihefte zu den Veröffentlichuwgen Naturschutz und Lanschaftspflege in Baden-Wurttemberg, Karlsruhe.* 21:35-47.
MORRIS, M.G. (1983a) Insects and the environment in the United Kingdom. *Atti del XII Congresso Nazionale Italiano di Entomologia,* Rome, 1980 203-235.
MORRIS, M.G. (1983b) Cashing in on the insect trade. *International Agricultural Development.* 3 (2):26-27.
MORRIS, M.G. (1986) The scientific basis of insect conservation. *Proceedings of the 3rd European Congress of Entomology, Amsterdam.* (part 3) 357-367.
MORRIS, M.G. (1987) Changing attitudes to nature conservation - the entomological perspective. *Biological Journal of the Linnean Society.* 32: 213-223.
MORRIS, M.G. (1989) Leglisation for Lepidoptera conservation - towards a rationale. *Nota Lepidopterologica,* 12 (Supplement 1): 15.
MORRIS, M.G. (1991) The management of reserves and protected areas. In: Spellerberg, I.F., Goldsmith, F.B. & Morris, M.G. *The Scientific Management of Temperate Communities for Conservation.* pp.323-347. 31st Symposium of the British Ecological Society. Blackwell Scientific Publications, Oxford.
MORRIS, M.G. (1992) Europe's butterflies: conserving a cultural and scientific resource. pp.4-14 In. Proceedings of International Congress Pavlicek-van Bek, T., Ovaa, A.H. & van der Made, J.G. (eds.) *Future of Butterflies in Europe: Strategies for Survival*, Wageningen, Netherlands.
MORRIS, M.G. & THOMAS, J.A.(1989) Re-establishment of insect populations, with special reference to butterflies (Chapter 2, *Moths and Butterflies of Great Britain and*

Bibliography

Ireland. 7 (2), ed. by J. Heath & A.M. Emmet, 22-36. Harley Books, Colchester, Essex.
MORRIS, M.G. & THOMAS, J.A. (1991) Progress in the conservation of butterflies. *Nota Lepidopterologica*, Suppl. 2, 32-44.
MORRIS, M.G. & WEBB, N.R. (1987) The importance of field margins for the conservation of insects. BCPC Monograph No. 35: *Field Margins.* 53-65.
MORRIS, M.G., COLLINS, N.M., VANE-WRIGHT, R.I. & WAAGE, J. (1991) *The utilisation and value of non-domesticated insects.* pp. 319-347. Proceedings of Royal Entomological Society of London, Symposium on Insect Conservation, London, September 1989.
MORRIS, M.G., THOMAS, J.A., WARD, L.K., SNAZELL, R.G., PYWELL, R.F., STEVENSON, M.J. & WEBB, N.R. Recreation of early successional stages for threatened butterflies - an ecological engineering approach. *Journal of Environmental Management.* (in press).
MORRISON, P. (1988) Conserving Britain's butterflies. *Environment Now.* July 30-32.
MORTON, A.C. (1982). The importance of farming butterflies. *New Scientist.* 20 May.
MORTON, A.C. (1979) Isolation as a factor responsible for the decline of the Large Blue butterfly (*Maculinea arion* (L.)) in Great Britain. *Entomologist's Monthly Magazine.* 115: 247-249.
MORTON, A.C. (1983) Butterfly Conservation - the need for a captive breeding institute. *Biological Conservation.* 25: (1) 19-33.
MUGGLETON, J. (1973) Some aspects of the history and ecology of blue butterflies in the Cotswolds. *Proceedings of the British Entomological and Natural History Society.* 6:77-84
MUGGLETON, J. (1975) Observations on *Lysandra coridon* Poda. (Lep., Lycaenidae) colonies at two sites in Gloucestershire using mark, release, recapture methods. *Proceedings of the British Entomological and Natural History Society.* 73-82.
MUGGLETON, J. & BENHAM, B.R. (1975) Isolation and the decline of the Large Blue butterfly (*Maculinea arion*) in Great Britain. *Biological Conservation.* 7: 119-128.
NATIONAL TRUST (1986) *Wicken Fen, An illustrated Souvenir.* The National Trust.
NATIONAL TRUST (1989) *2nd Review, July 1989, Papers to the Properties Committee.* Unpublished internal document.
NATIONAL TRUST (1992) *The National Trust and nature conservation.* leaflet. National Trust.
NATIONAL TRUST (1994) *Countryside Policy Review.* National Trust.
NATURE CONSERVANCY. Annual Reports.
NATURE CONSERVANCY COUNCIL, *Annual Reports*: 1st to March 1975 etc. i.e. 2nd to 1976, 3rd 1977, 4th 1978, 5th 1979, 6th 1980, 7th 1981, 8th 1982, 9th 1983, 10th 1984, 11th 1985, 12th 1986, 13th 1987, 14th 1988, 15th 1989, 16th (final) to 1990.
NATURE CONSERVANCY COUNCIL, (1977) *Nature Conservation and Agriculture.* Nature Conservancy Council, Peterborough.
NATURE CONSERVANCY COUNCIL, (1979) Press notice. Britain's rarest butterfly,

the Large Blue, probably extinct. 2pp.
NATURE CONSERVANCY COUNCIL, (1981) *The Conservation of Butterflies*. Nature Conservancy Council, Peterborough. 27pp.
NATURE CONSERVANCY COUNCIL, (1984) *Forestry Operations and Broadleaf Woodland Conservation. No. 8*. Nature Conservancy Council, Peterborough. 59pp.
NATURE CONSERVANCY COUNCIL, (1986) *Research and survey in nature conservation, No. 2 Monitoring the abundance of butterflies*. Nature Conservancy Council, Peterborough. 280pp.
NATURE CONSERVANCY COUNCIL, (1989) *Guidelines for selection of biological SSSIs, rationale, operational approach and criteria, detailed guidelines for habitats and species-groups*. Nature Conservancy Council, Peterborough. 288pp.
NATURE CONSERVANCY COUNCIL, (1990) *Internationally important bird sites: Special Protection Areas and Ramsar sites*. Nature Conservancy Council, Peterborough.
NEW, T. (1991) *The Conservation of Butterflies*. Oxford University Press, Melbourne.
NEW, T. & COLLINS, M. (1991) *Swallowtail Butterflies, An Action Plan for their Conservation*. IUCN - The World Conservation Union, Gland. 36pp.
NEWMAN, E. (1913) *Text Book of British Butterflies and Moths*.
NEWMAN, L.H. (1962) How our butterflies are dwindling: in the Isle of Wight, the Cotswolds, the Broads, and on the Downs, they are not where they used to be. *The Field*. 10 May.
NEWMAN, L.H. (1966) Butterflies on the Danger List. *Country Life*. 21 April p. 958-959.
NEWMAN, L.H. (UNDATED) *Butterfly Farmer*. Scientific Book Club Edition, London. see esp. Chapter 13 'To Introduce or not to Introduce' pp. 187-201.
OATES, M. (1993a) The Management of Southern Limestone Grasslands. *British Wildlife* 5:(2)73-82.
OATES, M.R. (1993b) Butterfly conservation in grassland habitats. Lecture presented at international symposium on the Ecology and Conservation of Butterflies organized by Butterfly Conservation. 10-12 September 1993 at Keele University, and (1995) as Butterfly conservation within the management of grassland habitats. In. *Ecology and Conservation of Butterflies*, Edited by A.S. Pullin. pp.98-112.
OATES, M.R. and WARREN, M.S. (1990) *A Review of Butterfly Introductions in Britain and Ireland*. World Wide Fund for Nature, Godalming. see WARREN, M. (1991) A Review of Butterfly Introductions in Britain and Ireland. World Wide Fund for Nature. London 1990. (cf. also review in *Antenna*, October 1991. Vol. 15).
OWEN, D.F. (1976) Conservation of butterflies in garden habitats. *Environmental Conservation*. 3:285-290.
OWEN, D. (1984) A Treasure house of butterflies. *Natural World*. Winter/Spring, 1983/4.8-9.
PAVLICEK-VAN BEEK, T. OVAA, A.H. & VAN DER MADE, J.G. (1992) Future of butterflies in Europe. *Proceedings of an International congress, held at Wageningen*

Bibliography

during April 12-15, 1989. Department of Nature Conservation, Agricultural University, Wageningen. 326pp.

PEACHEY, C.A. (1979) *The conservation of the Butterflies in Bernwood Forest*. Unpublished, Nature Conservancy Council.

PEACHEY, C.A. (1980) *The ecology of the butterfly communities of Bernwood Forest*. Unpublished MPhil Thesis. Oxford Polytechnic, Oxford.

PEACHEY, C.A. (1982) *National Butterfly Review, part 1: the representation of butterflies on National Nature Reserves. Invertebrate Site Register Report 10(1)*, Nature Conservancy Council, London. (Confidential unpublished report).

PETERKEN, G.F. (1977) General management principles for nature conservation in Britains woodlands. *Forestry* 50 (1) 27-48.

PETERKEN, G.F. & HARDING, P.T. (1974) Recent changes in the conservation value of woodlands in Rockingham Forest. *Forestry*. 47: 109-128.

PETERKEN, G.F. & HARDING, P.T. (1975) Woodland conservation in eastern England: comparing the effects of changes in three study areas since 1946. *Biological Conservation*. 8 279-298.

PHILLIPS, A.(1993) Butterflies and Nature Conservation. *Butterfly Conservation News*. (53) p.10-11.

PHILP, E. (1993) The Butterflies of Kent, an atlas of their distribution. *Transactions of the Kent Field Club*. Volume 12, Sittingbourne, 1993. 60pp.

PLANTLIFE, (1993) *The Acid Test for Plants. Acid rain and British Plants*. Plantlife. p.15.

POLLARD, E., (1979a) A national scheme for monitoring the abundance of butterflies: the first three years, *Proceedings of the British Entomological and Natural History Society*. 12: 77-90.

POLLARD, E. (1979b) Population ecology and change in range of the white admiral, *Ladoga camilla*, in England. *Ecological Entomology*, 4: 61-74.

POLLARD, E. (1982) Monitoring butterfly abundance in relation to the management of a nature reserve. *Biological Conservation*. 24: 317-328.

POLLARD, E. (1985a) Butterfly monitoring, 1984. Report to recorders. (CSD report no.567). Nature Conservancy Council, Peterborough.

POLLARD, E. (1985b) Butterfly monitoring sheme. (CSD report no.568) Nature Conservancy Council. Peterborough.

POLLARD, E. (1988) Temperature, rainfall and butterfly numbers. *Journal of Applied Ecology*. 25: 819-828.

POLLARD, E. & EVERSHAM, B.C. (1993) Butterfly monitoring 2 - interpreting the changes. Lecture presented at international symposium on the Ecology and Conservation of Butterflies organized by Butterfly Conservation. 10-12 September 1993 at Keele University.

POLLARD, E. & YATES, T.J. (1993) *Monitoring Butterflies for Ecology and Conservation*, The British Butterfly Monitoring Scheme. Published in association with

Bibliography

the Institute of Terrestrial Ecology (Natural Environment Research Council) and the Joint Nature Conservation. 256pp.

POLLARD, E., HALL, M.L., & BIBBY, T..J, (1986) *Monitoring the abundance of butterflies, 1976- 1985. Research and Survey in Nature Conservation. No. 2*, Nature Conservancy Council, Peterborough.

PORTER, K. (1993) Wide rides for butterflies. *Enact* Vol. 1., Number 1., Spring 1993, p. 17- 19.

PRATT, C. (1981) *A History of the Butterflies and Moths of Sussex, being a history and modern-day survey of the macrolepidoptera of East and West Sussex*. Borough of Brighton and the Booth Museum of Natural History. 356pp.

PRENDERGAST, J.R., QUINN, R.M., LAWTON, J.H., EVERSHAM, B.C. & GIBBONS, D.W. (1993) Rare species, the coincidence of deiversity hotspots and conservation strategies. *Nature*, 365:335-7.

PULLIN, A.S. (1995) (ed.) *Ecology and Conservation of Butterflies*. Chapman & Hall, London in association with The British Butterfly Conservation Society. 363pp.

PULLIN, A., McLEAN, I.F.G. & WEBB, M.R. (1993) The Large Copper : British and European perspectives. Lecture presented at international symposium on the Ecology and Conservation of Butterflies organized by Butterfly Conservation. 10-12 September 1993 at Keele University, and (1995) In. *Ecology and Conservation of Butterflies*, Edited by A.S. Pullin. pp.150-164.

PUREFOY, E.B. (1931) *Chrysophanus dispar batavus* Oberth., in Wicken Fen: the romance of a beautiful and very rare butterfly. *Entomologist*. 64: (823) 265-267.

PYLE, R.M. (1976) *The eco-geographic basis for Lepidoptera conservation. Part iii. A review of world Lepidoptera conservation*. Yale Univ. PhD. Thesis, Publication of the University Microfilm Int. 369pp.

PYLE, R.M. (1981) Lepidoptera conservation in Great Britain. *Atala*. 7: (2) 1979 (81) 34-43.

RAMSAR CONVENTION, see Convention on wetlands.

RANDS, M.R.W. & SOTHERTON, N.W. (1986) Pesticide use on cereal crops and changes in the abundance of butterflies on arable farmland in England. *Biological Conservation*. 36:71-82.

RATCLIFFE, D.A. (1967) Conservation and the collector. In. Duffey, E. *The Biotic Effects of Public Pressures on the Environment*. Monks Wood Experimental Station, Abbots Ripton.

RATCLIFFE, D. (1977) *A Nature Conservation Review*. Cambridge University Press, Cambridge. 2 vols.

RATCLIFFE, D. (1979) The end of the Large Blue butterfly. *New Scientist*. 84: 457-458.

RAVENSCROFT, N. (1986) The Silver-studded blue in Suffolk. *News of the British Butterfly Conservation Society*. 36: 35-36.

RAVENSCROFT, N. (1988) The Conservation of the Silver-studded blue in Suffolk. *News*. 40:18-20.

Bibliography

RAVENSCROFT, N. (1990) The ecology and conservation of the silver-studded blue butterfly *Plebejus argus* L., on the Sandlings of East Anglia, England. *Biological Conservation* 53:21-36.

RAVENSCROFT, N. (1992) *The ecology and conservation of the chequered skipper butterfly, Carterocephalus palaemon (Pallas)*. Ph.D. theis, University of Aberdeen.

RAVENSCROFT, N. (1993) The conservation of the chequered skipper butterfly in Scotland. Lecture presented at international symposium on the Ecology and Conservation of Butterflies organized by Butterfly Conservation. 10-12 September 1993 at Keele University, and (1995) In. *Ecology and Conservation of Butterflies*, Edited by A.S. Pullin. pp.165-179.

RAVENSCROFT, N.O.M. & WARREN, M.S. (1992) Habitat selection by larvae of the chequered skipper *Carterocephalus palaemon* (Pallas) (Lepidoptera: Hesperidae) in northern Europe. *Entomologist's Gazette* 43: 237-242.

READ, M. (1985) *The silver-studded blue conservation report*. M.Sc thesis, University of London.

REED, T.M. (1982) The number of butterfly species on British Islands. *Proceedings of 3rd European Lepidopterology, Cambridge* 1982 146-152.

RENNIE, J. (1832) *A Conspectus of the Butterflies and Moths found in Britain*. London.

RILEY, N.D. (1929) The re-establishment of the Large Copper butterfly (*Chrysophanus dispar*) in England. *Natural History Magazine*.12 (11)113-118.

RIPLEY, I. (1993) A Provisional Atlas of the Butterflies of Northern Ireland. 28pp.

ROBERTSON, P.A., CLARKE, S.A. & WARREN, M.S. (1993) Woodland management and butterfly diversity. Lecture presented at international symposium on the Ecology and Conservation of Butterflies organized by Butterfly Conservation. 10-12 September 1993 at Keele University, and (1995) In. *Ecology and Conservation of Butterflies*, Edited by A.S. Pullin. pp.113-122..

ROBERTSON, T.S. (1980) An estimate of the British population of *Apatura iris* (Linneaus). (Lepidoptera: Nymphalidae.). *Proceedings of the British Entomological and Natural History Society*. pp.89-94.

ROBERTSON, T.S. (1981) The decline of *Carterocephalus palaemon* (Pallas) and *Maculinea arion* (L.) in Great Britain. *Entomologist's Gazette*. 32: 5-12.

ROBINSON, H.S. (1952) Some suggestions on the examination of an ethical and practical problem. *Entomologist's Gazette* 3:45-51.

ROTHSCHILD, M. (1946) *Letter* dated 29 November 1946 to Captain Riley In: JCCBI archives at RESL.

ROTHSCHILD, M. (1975) The Swallowtail Butterfly *Papilio machaon britannicus* Seitz in Northamptonshire. *Entomologist's Record and Journal of Variation.* 87:178-179.

ROTHSCHILD, M. (1979) Nathaniel Charles Rothschild 1877-1923.p.3-12. Privately printed by Miriam Rothschild, Cambridge University Press. Cambridge.

ROTHSCHILD, M. (1983) *Dear Lord Rothschild, birds, butterflies & history.*

Bibliography

Hutchinson, London. 398pp.

ROTHSCHILD, M. (1987) Changing conditions and conservation at Ashton Wold, the birthplace of the SPNR. *Biological Journal of the Linnean Society* 32:161-170.

ROTHSCHILD, M. (1989) Ashton Wold, some changes in the Fauna and Flora between 1900 and 1989. p. 29-38 In. The Nature of Northamptonshire. Edited by Adrian Colston and Franklyn Perring. Northamptonshire Wildlife Trust.

ROTHSCHILD, M. (1992) The Silver-washed Fritillary (*Argynnis paphia*) to Ashton Wold. *Antenna* 16:(4)134.

ROTHSCHILD, M. & FARRELL, C. (1983) *The Butterfly Gardener*. Michael Joseph / Rainbird, London. 128pp.

ROTHSCHILD, N.C. (1915a) Dutch *Chrysophanus dispar*. *Proceedings of the Entomological Society of London* III, IV, p.lxxx.

ROTHSCHILD, N.C. (1915b) *Society for the Protection of Nature Reserves. Schedule of Areas of England, Scotland, Wales and Ireland.* Private and Confidential, Privately published.

ROWELL, T.A. (1987) History and experimentation in the management of Wicken Fen. *Nature in Cambridgeshire* (28) 14-19.

ROWELL, T.A. (1991) *Wildlife Link, SSSIs: A Health Check, a review of the statutory protection afforded to sites of special scientific interest in Great Britain.* Wildlife Link, London. 52pp.

ROWELL, T.A. & HARVEY, H.J. (1988) The recent history of Wicken Fen, Cambridgeshire, England. A guide to Ecological Development. *Journal of Ecology.* (1988) 76: 73-90.

ROWELL, T.A., GUARINO, L. & HARVEY, H.J. (1985) The experimental management of vegetation at Wicken Fen, Cambridgeshire. *Journal of Applied Ecology* 22:217-227.

ROYAL ENTOMOLOGICAL SOCIETY OF LONDON (1957) *Confidential List of rare localised British Insects. 3rd December, 1957.*

ROYAL SOCIETY FOR THE PROTECTION OF BIRDS (1994) *Biodiversity Challenge: an agenda for conservation action in the UK.* pp.137. A consultative document prepared by a group of voluntary conservation organisations: Butterfly Conservation, Friends of the Earth, Plantlife, The Royal Society for Nature Conservation - the Wildlife Trusts' partnership, The Royal Society for the Protection of Birds and World Wide Fund for Nature. cf. Wynne *et al.* (1995).

RUSSELL, S.G.C. (1925) The causes of scarcity of *Polyommatus thetis* and *coridon*. *Entomologist* 58:100-1.

SAWFORD, B. (1987) *The Butterflies of Hertfordshire*. Castlemead Publications, Ware. 195pp. see Chapter 10: Conservation and the Future for the Butterflies of Hertfordshire. p.168-176.

SHELDON, W.G. (1925) A committee for the protection of British Lepidoptera. *Entomologist* 58:153-154.

Bibliography

SHEPPARD, D.A. (1990a) *Notes on the management of Trislington Plantations for invertebrates.* England Field Unit Project no. 22. 19pp.

SHEPPARD, D.A. (1990b) Changes in the Fauna of magnesian limestone grassland after transplantation: preliminary observations at Thrislington Plantations. In. *Calcareous Grasslands - Ecology and Management.* Proceedings of a joint British Ecological Society/Nature Conservancy Council symposium, 14-16 September 1987 at the University of Sheffield. 171-172.

SHIRT, D.B. (ed.) (1987) *British Red Data Books: 2. Insects.* Nature Conservancy Council, Peterborough. 402pp.

SHOARD, M. (1980) *The Theft of the Countryside.* Temple Smith, London. 271pp.

SHOARD, M. (1987) *This Land is Our Land, The Stuggle for Britain's Countryside.* Paladin, London. 592pp.

SHREEVE, T. (1993) The mobility of the British Butterfly Fauna. Lecture presented at international symposium on the Ecology and Conservation of Butterflies organized by Butterfly Conservation. 10-12 September 1993 at Keele University.

SOCIETY FOR THE PROMOTION OF NATURE RESERVES (1923 onwards) *Handbook of the Society for the Promotion of Nature Reserves, Lincoln.*

SOCIETY FOR THE PROMOTION OF NATURE RESERVES (1943) Memorandum No. 3. *Nature Conservation in Great Britain.* Issued following the Conference on Nature Preservation in Post-War Reconstruction.

SOCIETY FOR THE PROMOTION OF NATURE RESERVES (1970) *A Policy on Introduction to Nature Reserves.* S.P.N.R. Conservation Liaison Committee, Technical Publication No. 2. Published by the S.P.N.R. by The Association of Nature Conservation Trusts. 14pp.

SOMMERVILLE, A. (1977) The conservation of the chequered skipper butterfly. *Scottish Wildlife.* 13:24-25; 49.

SOUTH, R. (1884) On the probable extinction of *Lycaena arion* in England. *Entomologist's Monthly Magazine.* 21:133-134.

SOUTHWOOD, T.R.E. (1951) Proceedings of the Royal Entomological Society of London (C) 16:23-4.

SOUTHWOOD, T.R.E. (1965) *Insect Conservation: An Appeal.* 2pp leaflet. Royal Entomological Society.

SOUTHWOOD, T.R.E. (1975) *Ecological Methods, with particular reference to the study of insect populations.* Chapman & Hall, London. 391pp.

SPOONER, G.M. (1946) *Confidential report.* 6pp. In: Minutes of the Committee for the Protection of the British Lepidoptera.

SPOONER, G.M. (1963) On causes of the decline of *Maculinea arion* L. (Lep. Lycaenidae) in Britain. *Entomologist.* 96: 199-210.

STAINTON, H.T. (1857) *A Manual of British Butterflies and Moths.* Volume 1. London, John Van Voorst.

STAMP, Sir D. (1969) *Nature Conservation in Britain.* Collins, London. New Naturalist

Bibliography

Series. 273pp.

STEEL, C. (1991) *Woodland Butterflies*. British Butterfly Conservation Society. 48pp

STEEL, C. & STEEL, D (1985) *Butterflies of Berkshire, Buckinghamshire and Oxfordshire*. Pisces Publication, Oxford. pp.80. See section on Conservation, p.12-14.

STEELE, R.C. (1981) Nature conservation in Britain. In. *Nature conservation and recreation in European heritage landscapes*. edited by A.J. Youel, 44-53. Castleton, Peak Park Joint Planning Study Centre.

STEELE, R.C. (1983) Management of woodland and woodland vegetation for wildlife conservation. In *Management of vegetation*. ed. by J.M. Way, 131-143. British Crop Protection Council, Croydon.

STUBBS, A.E. (1981a) *The conservation of butterflies*. Nature Conservancy Council, Peterborough.

STUBBS, A.E. (1981b) Invertebrate conservation in the Nature Conservancy Council Great Britain. *Atala*. 7 (2) 58-60.

STUBBS, A.E. (1981c) Loss of Lepidoptera habitat in Britain: part of a European problem. *Veröffentlichungen für Naturschutz und Landschaftspflege in Baden-Wurtenberg* 21: 49-58.

STUBBS, D. (1988) *Towards an introductions policy: Conservation guidelines for the introduction and reintroduction of living organisms into the wild in Great Britain*. Pisces Publications, Oxford.

STUBBS, A.E. (1991) Protected British Buterflies:interpretation of Section 9 and Schedule 5 of the Wildlife and Countryside Act 1981. *Entomologist's Record and Journal of Variation*. 103:197-199

STUBBS, A. (1994) Our role in the UK Biodiversity Initiative, maintaining the biodiversity of butterflies and moths in the UK. *Butterfly Conservation Annual Review 1993*, Number 2. 5-7.

SUTTON, S.L. & BEAUMONT, H.E. (1989) *Butterflies and Moths of Yorkshire, distribution and conservation*. Yorkshire Naturalist's Union. pp.367. See Chapter 4: Conservation of Sites and Conservation of Species.

THOMAS, C.D. (1993) The implications of medium and large scale population processes for butterfly conservation. Lecture presented at international symposium on the Ecology and Conservation of Butterflies organized by Butterfly Conservation. 10-12 September 1993 at Keele University.

THOMAS, C.D. (1995) Ecology and conservation of butterfly metapopulations in the fragmented British landscape. In. *Ecology and Conservation of Butterflies*, Edited by A.S. Pullin. pp.46-63.

THOMAS, C.D. & JONES, T.M. (1993) Partial recovery of a skipper butterfly (*Hesperia comma*) from poipulation refuges: lessons for conservation in a fragmented landscape. *Journal of Animal Ecology*. 62:472-81.

THOMAS, C.D. THOMAS, J.A. & WARREN, M.S. (1992) Distributions of occupied and vacant butterfly habitats in fragmented landscapes. *Oecologica*. 92:563-567.

Bibliography

THOMAS, J.A. (1974) Factors influencing the numbers and distribution of the Brown hairstreak, *Thecla betulae* L. (Lepidoptera,Lycaenidae) and the Black hairstreak, *Strymonidia pruni* L. (Lepidoptera, Lycaenidae). PhD. Thesis, Leicester University. 228pp.

THOMAS, J.A. (1975) *The black hairstreak, conservation report*. Unpublished report ITE/NCC.

THOMAS, J.A. (1976a) The biology and conservation of the Large Blue butterfly *Maculinea arion* L. *ITE contract report to Nature Conservancy Council.*

THOMAS, J.A. *(1976b) The Black Hairstreak - conservation report.* Nature Conservancy Council, Banbury. 58pp.

THOMAS, J.A. (1977a) Ecology and conservation of the Large Blue butterfly - second report. *ITE contract report to Nature Conservancy Council. Institute of Terrestrial Ecology Report number 400. 23pp.*

THOMAS, J.A. (1977b) *Proceedings of the Royal Entomological Society.* 16.2.77.

THOMAS, J.A. (1977c) The ecology of the Large Blue butterfly. *Annual Report of the Institute of Terrestrial Ecology.* 1976 25-27.

THOMAS, J.A. (1978) *Report on the Large Blue butterfly in 1978*. Nature Conservancy Council, Institute of Terrestrial Ecology, Furzebrook, Wareham. Nature Conservancy Council, Institute of Terrestrial Ecology project 400. 30pp.

THOMAS, J.A. (1980a) The extinction of the Large Blue and the conservation of the Black Hairstreak (a contrast of failure and success). *Report of the Institute of Terrestrial Ecology.* 1979 19-23.

THOMAS, J.A. (1980b) *The Glanville fritillary; survey 1979*. Joint Committee for the Conservation of British Insects, London. (with Simcox, D.J.) 37pp.

THOMAS, J.A. (1980c) Why did the Large Blue become extinct in Britain? *Oryx* 15: 243-247.

THOMAS, J.A. (1981a) Butterfly year, 1981-2. *Atala* 7, (1979) 52.

THOMAS, J.A. (1981b) Insect conservation in Britain: some case histories. *Atala*, 6, (1978) 31-36.

THOMAS, J.A. (1983) The ecology and conservation of *Lysandra bellargus* (Lepidoptera:Lycaenidae) in Britain. *Journal of Applied Ecology* 20: 59-83.

THOMAS, J.A. (1984) The conservation of butterflies in temperate countries: past efforts and lessons for the future. In. *The Biology of Butterflies*. Symposium of the Royal Entomological Society of London. Number 11. pp. 333- 351 Academic Press, London.

THOMAS, J.A. (1985) The status and conservation of the Buterfly *Plebejus argus* L. (Lepidoptera:Lycaenidae) in North West Britain. *Biological Conservation.* 33:(1985)29-51.

THOMAS, J.A. (1986) *RSNC Guide to Butterflies of the British Isles.* Country Life Books, an imprint of Newnes Books, a division of Hamlyn, London. 160 pp.

THOMAS, J.A. (1987) Introductions to conserve the Adonis blue butterfly. p.41,95-97. In: *The Natural Environment Research Council, Report for the Council for the period 1*

Bibliography

April 1986 to 31 March 1987, Institute of Terrestrial Ecology.
THOMAS, J.A. (1989a) The return of the large blue butterfly. *British Wildlife.* 1 (1) 2-13.
THOMAS, J.A. (1989b) The history, decline and re-establishment of the large blue butterfly in Devon. *Nature in Devon.* 10: 34-43.
THOMAS, J.A. (1989c) Ecological lessons from the re-introduction of Lepidoptera. *Entomologist.* 108: (1-2) pp. 56-68.
THOMAS, J.A.T. (1990) The Large blue butterfly in 1989/90. *Report to the World Wide Fund for Nature. March 1990.*
THOMAS, J.A. (1991) Rare species conservation: case studies of European butterflies. In. Spellerberg, I.F., Goldsmith, F.B. & Morris, M.G., 1991. *The Scientific Management of Temperate Communities for Conservation.* Published for the British Ecological Society by Blackwell Scientific Publishers, Oxford. pp. 566.
THOMAS, J.A. (1994) Lecture on 'Conservation of Insects' delivered at The Royal Entomological Society, 16 February 1994.
THOMAS, J.A. (1995) The ecology and conservation of *Maculinea arion* and other European species of large blue butterfly. In. *Ecology and Conservation of Butterflies,* Edited by A.S. Pullin. pp.180-197.
THOMAS, J.A. & COLLINS, M. (1991) *Conservation of Insects and their habitats.* Academic Press, London.
THOMAS, J.A. & MORRIS, M.G. (1986) Proposals for a Federation of Butterfly Farmers. *Butterfly News.* 9: 6.
THOMAS, J.A. & WEBB, N. (1984) *Butterflies of Dorset.* Dorset Natural History & Archaelogical Society, Dorcester. 128pp.
THOMAS, J.A., THOMAS, C.D., SIMCOX, D.J., CLARKE, R.T. (1986) The ecology and declining status of the silver-spotted skipper butterfly (*Hesperia comma*) in Britain. *Journal of Appied Ecology* 23:365-380.
THOMSON, G. (1976) Our 'disappearing' butterflies. *Forth Naturalist and Historian.* 1: 89-105.
THOMSON, G. (1980) *The Butterflies of Scotland, A Natural History.* Croom Helm, London. 267 pp.
THOMSON, G (1990) *European Butterfly Identifier, a computerised identification system and database.* Computer software providing instant access to ecological data. ID Software, Lockerbie.
TOMLINSON, D. (1992) High noon for the high brown: future of a rare fritillary. *Country Life.* 20 August p. 65.
TUDOR, O. (1976) Butterfly conservation in woodlands. *Quarterly Journal of Forestry.* 70 (2) 92-95.
TUDOR, O. (1979) Butterfly conservation in woolands - further comments. *Quarterly Journal of Forestry.* 73:157-160.
Van SCHEPDAEL, J. (1962) Genese du peuplement de *Thersamonia dispar dispar*

Bibliography

Haworth en Angleterre et *Th. dispar batavus* Oberth. en Frise. *Linneana belgica*. 1: 14-27.

VANE-WRIGHT, R.I. (1977) Two approaches to insect conservation. Reports in *Antenna* July 1977 (1) p. 17.

VANE-WRIGHT, R.I. (1978) Ecological and behavioural origins of diversity in butterflies. In. *Diversity of Insect Faunas*. Edited by L.A. Mound and N. Waloff. Symposium of the Royal Entomological Society of London: No. 9. Chapter 4. Academic Press, London.

VANE-WRIGHT, R. & ACKERY, P. (1990). *The Biology of Butterflies*. Symposium of the Royal Entomological Society of London. No. 11. Academic Press, London. 429pp.

VANE-WRIGHT, R.I., HUMPHRIES, C.J. & WILLIAMS, P.H. (1991) What to protect? - systematics and the agony of choice. *Biological Conservation*. 55: 235-254.

VERRALL, G.H. (1909) The 'large copper' butterfly (*Chrysophanus dispar*) *Entomologist* 42:183.

VICKERY, M.(1993) The value of gardens to butterflies. Lecture presented at international symposium on the Ecology and Conservation of Butterflies organized by Butterfly Conservation. 10-12 September 1993 at Keele University, and (1995) as Gardens: the neglected habitat. In. *Ecology and Conservation of Butterflies*, Edited by A.S. Pullin. pp.123-134.

WALKER, A.M.(1983) Entomology and Conservation:The Importance of MOD Land. *Sanctuary, Conservation Bulletin MOD* No. 11: 29-30.

WARREN, A. & GOLDSMITH, F.B. (eds.) (1974) *Conservation in Practice*. Wiley, London.

WARREN, M.S. (1981a) The Ecology of the wood white butterfly *Leptidea sinapis* (L.) (Lepidoptera, Pieridae). Ph.D. Thesis, University of Cambridge.

WARREN, M.S. (1981b) *The heath fritillary, Mellicta athalia, in 1981*. Report to the Joint Committee for the Conservation of British Insects. 16pp.

WARREN, M.S. (1983) The ecology and conservation of the heath fritillary butterfly, *Mellicta athalia* Rott., in Britain. *Biological Conservation* 29:287-305.

WARREN, M.S. (1984a) Conserving the heath fritillary. *Ecos* 5(1):11-13.

WARREN, M.S. (1984b) The future of the heath fritillary butterfly in Britain. *News of the British Butterfly Conservation Society*. No. 32: 19-30.

WARREN, M.S. (1984c) The heath fritillary butterfly in 1983. *Annual Report of the Vincent Wildlife Trust*. 1983:25-26.

WARREN, M.S. (1984d) The biology and status of the wood white butterfly, *Leptidea sinapis* (L.) (Lepidoptera, Pieridae) in the British Isles. *Entomologist's Gazette*. 35 207-223.

WARREN, M.S. (1984e) The status of the Heath Fritillary in Britain. *Biological Conservation*. 29:(4) 287-305.

WARREN, M.S. (1985a) The status of the heath fritillary buterfly, *Mellicta athalia*, in relation to changing woodland management in the Blean Woods, Kent. *Quarterly*

Bibliography

Journal of Forestry. 79: 174-182.

WARREN, M.S. (1985b) The influence of shade on butterfly numbers in woodland rides, with special reference to the wood white, *Leptidea sinapis. Biological Conservation.* 33:147-164.

WARREN, M.S. (1986) Notes on habitat selection and the larval host-plants of the brown argus, *Aricia agestis,* marsh fritillary, *Eurodryas aurinia,* and painted lady, *Vanessa cardui,* in 1985. *Entomologist's Gazette.* 37:65-67.

WARREN, M.S. (1987a) Conserving Berkshires rarer butterflies. *Berkshire Matters* No. 17, 4-5. Berks Bucks & Oxon Naturalists' trust, Oxford.

WARREN, M.S. (1987b) The ecology and conservation of the heath fritillary butterfly, *Mellicta athalia* L. I. Host selection and phenology. II. Adult population structure and mobility. III, Population dynamics and the effect of habitat management. *Journal of Applied Ecology.* 24: 467-482, 483- 498, 499-513.

WARREN, M.S. (1988) Supporting butterfly conservation research. In: Stansfield, G. & Harding, P.T. (eds.) *Biological recording: the products.* pp. 40-50. National Federation for Biological Recording, Cambridge.

WARREN, M.S. (1989a) *Mellicta athalia* Rott.: An example of successful Lepidoptera conservation in the United Kingdom. *Nota Lepidopterologica,* supplement No. 1: 21-22.

WARREN, M.S. (1989b) Pheasants and fritillaries: Is there really any evidence that pheasant rearing may have caused butterfly declines? *British Journal of Entomology and Natural History.* 2:169-175.

WARREN, M.S. (1990a) The conservation of *Eurodryas aurinia* in the UK. In: *Colloquy on the Berne Convention Invertebrates and their conservation: conclusions and summaries,* pp.71-74. Council for Europe, Strasbourg.

WARREN, M.S. (1990b) Draft data sheet - European Invertebrate Survey: *Euphydras (Eurodryas) aurinia.* In. *Convention on the Conservation of European Wildlife and Natural Habitats,* Standing Committee 10th meeting, Strasbourg, 8-11 January 1990. Council of Europe, Strasbourg.

WARREN, M.S. (1990c) European butterflies on the brink. *British Wildlife.* 1(4):185-196.

WARREN, M.S. (1991a) The successful conservation of an endangered species, *Mellicta athalia* (the heath fritillary butterfly), in Britain. *Biological Conservation.* 55:37-56.

WARREN, M.S. (1991b) Woodland edge management for butterflies. In: Ferris-Kaan, R. (Ed). *Edge management in woodlands.* FC Occasional paper No. 28, pp. 22-24. Forestry Commission, Farnham.

WARREN, M.S. (1991c) Bringing the chequered skipper back to England: a study of its habitats in northern Europe. *Butterfly Conservation News.* 47: 47-51.

WARREN, M.S. (1991d) The chequered skipper, *Carterocephalus palaemon, in northern Europe.* British Butterfly Conservation Society, Occasional Paper No. 2.

WARREN, M. (1992a) Britain's Vanishing Fritillaries. *British Wildlife.* 3: (5) 282- 296.

Bibliography

WARREN, M.S. (1992b) Butterfly populations. In. Dennis, R.L.H. (Ed) *The Ecology of Butterflies in Britain*. 73-92. Oxford University Press, Oxford.

WARREN, M.S. (1992c) The conservation of British butterflies. In. Dennis, R.L.H. (ed.) *The Ecology of Butterflies in Britain*. 246-274. Oxford University Press, Oxford.

WARREN, M.S. (1992d) Conservation research on *Mellicta athalia*, an endangered species in the U.K. In: Pavlicek-van Beek, T., Ovaa, A.H. & Van der Made, J.G. (1992) *Future of butterflies in Europe*. Proceedings of an International congress, held at Wageningen during April 12-15, 1989. Department of Nature Conservation, Agricultural University, Wageningen. 326 pp. pp. 124-133.

WARREN, M.S. (1992e) The high brown fritillary - Britain's most endangered butterfly? *Butterfly Conservation News*. 50:26-30.

WARREN, M.S. (1993a) A review of butterfly conservation in central southern Britain. I. Protection, evaluation and extinction on prime sites. *Biological Conservation*. 64:25-35.

WARREN, M.S. (1993b) A review of butterfly conservation in central southern Britain. II. Site management and habitat selection of key species. *Biological Conservation*. 64:37-49

WARREN, M.S. (1993c) Managing local microclimates for the high brown fritillary, *Argynnis adippe*. Lecture presented at international symposium on the Ecology and Conservation of Butterflies organized by Butterfly Conservation. 10-12 September 1993 at Keele University, and (1995) In. *Ecology and Conservation of Butterflies*, Edited by A.S. Pullin. pp.198-210.

WARREN, M.S. (1994) Making a mark for butterflies. *Butterfly Conservation Annual Review 1993*. Number 2. p.8-9.

WARREN, M.S. (1995) The UK status and suspected metapopulation structure of a threatened European butterfly, *Eurodryas aurinia* (the marsh fritillary). *Biological Conservation*. 67:239-249.

WARREN, M.S. (1995b) Habitat selection by the butterfly *Eurodryas aurinia*. in press.

WARREN, M.S. & FULLER, R.J. (1990) *Woodland rides and glades - their management for wildlife*. Nature Conservancy Council, Peterborough. 32pp. Republished in 1992 as 2nd edition, published by JNCC.

WARREN, M.S. & KEY, R.S. (1991) Woodlands:past, present and potential for insects. In: Collins, N.M. & Thomas, J.A. (eds.). *The conservation of insects and their habitats*, 155-211. Academic Press, London.

WARREN, M.S. & LAWSON, T. (1990) Declining butterflies. *Green Magazine*. July, 1990: 185-196.

WARREN, M.S. & STEVENS, D.E.A. (1989) Habitat design and management for butterflies. *Entomologist*. 108: (1-2) 123-134.

WARREN, M.S. & THOMAS, J.A. (1992) Butterfly responses to coppicing. In: Buckley, G.P. (Ed) *The ecological effects of coppice management*, 249-270. Chapman & Hall,

Bibliography

London.
WARREN, M.S. & THOMAS, J.A. (1993) Conserving the silver-spotted skipper in practice. *Butterfly Conservation News.* 54:21-26.
WARREN, M.S. & THOMAS, J.A. (in press) Management options for the silver-spotted skipper: a study of the effects of timing of grazing at Old Winchester Hill NNR, Hants. *English Nature Science*
WARREN, M.S., THOMAS, C.D. & THOMAS, J.A. (1981) *The heath fritillary: survey and conservation report* (1980). Joint Committee for the Conservation of British Insects, London. 90pp.
WARREN, M.S., POLLARD, E. & BIBBY, T.J. (1986) Annual and long-term changes in a population of the wood white butterfly, *Leptidea sinapis*. *Journal of Animal Ecology.* 55: 707-719.
WARREN, M.S. MUNGUIRA, M.L. & FERRIN, J. (1994) Notes on the distribution, habitats and conservation of *Eurodryas aurinia* (Rottemburg) (Lepidoptera: Nymphalidae). *Entomologist's Gazette.* 44: *45:5-12.*
WATSON, J.N.P. 1977. The World's First Nature Reserve? *Country Life. 29* September pp. 844-845.
WEBB, N.R. & HOPKINS, P.J. (1984) Invertebrate diversity on fragmented *Calluna* heathland. *Journal of Applied Ecology* 21:921-933.
WELCH, R.C. (1993) After the split-entomology in the new conservation agencies. Royal Entomological Society's East Region Meeting, Peterborough, 28 October 1992, *Antenna,* p.18-19.
WELLS, T.C.E. (1969) Botanical Aspects of Conservation Management of Chalk Grasslands. *Biological Conservation.* 2: (1) 36-44.
WHALLEY, P. (1980) *Butterfly Watching.* Severn House Naturalist's Library, London. 160pp.
WILKINSON, W. (1990) Nature conservation in 1990. Presentation by the chairman of the Nature Conservancy Council, Sir William Wilkinson for the launch of the report for 1989 on 28 November 1990. Lecture transcript.
WYNNE, G., AVERY, M., CAMPBELL, L., GUBBAY, S., HAWKSWELL, S., JUNIPER, T., KING, M., NEWBERRY, P., SMART, J., STEEL, C., STONES, T., STUBBS, A., TAYLOR, J., TYDEMAN, C., and WYNDE, R., (1995) *Biodiversity Challenge, an agenda for conservation action in the UK.* Second Edition. 285pp. (NB The RSPB at Sandy, Beds. is acting as secretariat for this publication.)
WILLIAMS, M.J. (1989/90) Threat to Large Heath. BBCS West Midlands branch. *Newsletter No. 22 / Peat Campaign. BBCS West Midlands branch Newsletter. no. 23.*
WILLMOTT, K.J. (1987) The Ecology and Conservation of the Purple Emperor Butterfly *(Apatura iris).* Report on project BSP/2 for World Wildife Fund 1984-1986, sponsored by Associated Tyre Specialists, Harrow.
WILLMOTT, K. (1991a) *The Purple Emperor Butterfly.* British Butterfly Conservation Society. 20pp.

Bibliography

WILLMOTT, K. (1991b) The adaptation to a hostile environment by changing ovipositing cues by females of the Silver-spotted skipper (*Hesperia comma* Linn.1758). *Entomologist's Record and Journal of Variation.* p. 247-248.

WINOKUR, L. (1992) Climatic influence on Lysandra butterflies (Lepidoptera: Lycaenidae): The case for captive breeding. *Entomologist's Gazette.* 1992 43: (3) 163-165.

WORKING GROUP ON INTRODUCTIONS OF THE UK COMMITTEE FOR INTERNATIONAL NATURE CONSERVATION (1979) *Wildlife Introductions to Great Britain, the introduction, re-introduction and restocking of species in Great Britain: some policy implications for nature conservation.* 32pp.

YOUNG, M.R. & RAVENSCROFT, N. (1991) Conservation of the chequered skipper butterfly (*Carterocephalus palaemon* Pallas) in Scotland. Confidential report to Nature Conservancy Council.

INDEX

Acer pseudoplatanus 103
acid rain 124
Adonis Blue 98,105,107,114,122, 141,159
Adventurer's Fen 15
AES code 160
Afton Down SSSI 118
agro-chemicals 123
Alnus glutinosa 58
Aporia crataegi 159
Arne 16
Ashdown Forest 16
Ashton Wold 29
Associated Tyres 102
autecology 99

Balmoral 108
Barker, L & A 104
Barnwell Wold 88
Bath White 3
BBCS 16
BCCPN 12
Berger's Clouded Y. 155
Bern Convention 134
Bernwood Forest 29,40,100,109
Berry, Professor Sam 8,95
Betula sp. 58
Bill Smyllie reserve 166
Biodiversity 16,30,32,34,44
Bixley Farm 122
Black hairstreak 33,99,100, 101,109,159
Black-veined White 13
Blean, 146
BMS 20,30,146,165,166
Bonn Convention 130
Bourne Bottom 142
box junctions 100
Brachypodium pinnatum 101
Brassica napa 123
Braunton Burrows 16
Brecon 166
Brimstone Project 42
British Coal 117
British islands 2
Broads, The 51,60
Brown Argus 104,122
Brown Hairstreak 103,155
browns 5,108
BTO 20
bulldozers 105
BUTT 104
Butterfly Conservation 2,5165,100, 104,108,117,159,164,156,166,
Butterfly Conservation codes 161-162,188,190
Butterfly Conservation reserves 166-167
butterfly houses 131
Butterfly Valley 73
Butterfly Year 42

Cadbury, J.C. 79
calcareous grassland 104
Calluna vulgaris 120
Camberwell Beauty 3
Canford Heath 142
captive breeding 60
Castor Hanglands 42,146
Catfield Fen 54
CCCPN 12
CCRESL 14
Chalkhill Blue 98,107,141
Channel Tunnel 122
Chapman, T.A. 90
Chartwell 159
Chequered Skipper 19,27,29, 33,40,42,78,141,144
Chilterns 105,108
Chippenham Fen 147
Churchill, W. 159
Cinxia hunt 155
Cistus sp. 120
CITES 131
Clarke, Cyril 72,87
Clayden, C.N. 37
climax communities 36
clines 85
Cocayne, E.A. 79
Code for Dealers 179
Code for Insect Collecting 175
Code of Collecting 157
codes 153
collecting 35,56,83,157
Collier, Ray 43
colour forms 46
Comma 29
Common Blue 29,40,42, 122,165
Conservation committees 13
Continental influence 3
Copper Fields 60
Cornwall 165
Cotswolds 119
Cow-parsley 46
CPBL 12,14,90
CPRE 14
Croucher, Peter 150
Crown Estate Commission 115
Cruciferae 5
CRWS 14

Dark Green fritillaries 102
Darwin Initiative 32
Dempster, Jack 48-49
Devon 80-81
Dig for Victory 15
Dingy Skipper 29
Diver, C. 15
Diversity 28

Dizzard, The 73-75,80,96,155
DNA studies 68
Dorset RDBs 139
downland 105,108
Downs 122
drainage 65
Drake, Martin 150
Duffey, Eric 48,49
Duke of Burgundy 7,29,141,155
Duke of Cornwall 165
Dungeness SSSI 14,133
Durham 166

EC Directives 133
Edelsten, H.M. 13,25
Edwards, T.G. 75
Ellis, Ted 51,54
English Nature 115,119,125, 150
ENCY '95 42
English Nature 17,21,92
Entomological Society 12
Erica cinerea 120
Eriophorum vaginatum 143
ESA 133
EU 130

Farn, A.B. 159
Farrell, Lynne 43
FC 103,106,115
Fearnehough, T.D. 54
Festival of Britain 41,155
fines 141
flagship species 39
Flora for Fauna 42
FNR 31,125,164,166
FOE 32
Folkestone Warren 122
Ford, E.B. 15,35,50,85,88
Forest of Dean 159
Forestry Commission 35
Frankland, Thomas 16
fritillaries 129,141
Frohawk, F.W. 72,90
Fryer, J.C.F. 13,15

GAP 130
Gardiner, Brian 48,49
Gatekeeper 30
GATT 130
genetic effects 52,84
Gibbs, Julian 16
glades 100
Glanville Fritillary 13,33,75, 115,140,155
global warming 119
Goodden, R & R 87
Grass Wood, 98
Grasslands 118-119
Grayling 105,132,143
grazing 61,100,104,105,108

Green Hairstreak 29
Green-veined White 165
Grizzled Skipper 29
Guernsey 42

habitat change 76
habitat conservation 18
habitat loss 114
habitat management 97
habitat mosaics 36
Habitat Teams 17
Habitats Directive 116-117,134-136
Hall, Marney 48,102
Ham Street, 146
Hamilton, Duchess of 42
Harris, Moses 102
Harrison, Jeffery 68
Hartland Moor 16
Heath Fritillary 7,13,19,20,33,40, 98,99,101,107,140,141,155
heathlands 120-121
hedgerows 122
Hedges, A.H. 74,78,79
Hickling Broad 49,165
High Brown Fritillary 33,99,114, 140,147
Hillingdon, Lord 30
Hipparchia semele thyone 140
Hippocrepis comosa 105
Historical background 11
Holland & Barrett 92
Holly Blue 30
Horseshoe vetch 105
Howarth, Graham 75

immigration 3
Insect re-establishment 181
international laws 132
Isle of Purbeck 15,16
Isolation 52
ITE 16,18,72,80.87.95
IUCN 14,16,32,138

Jackson, R.A. 73
JCCBI 14,32,42,123
JCCLBB 80
JNCC 136
Jordan, Karl 55
Joy, Jenny 143

Keepers 157
Kent 166
Kettlewell, H.B.D. 95
keystone species 39,65
Kingley Vale 16
Kirkland, Paul 117
Kudrna, Otakar 7

Labouchere, A.J. 59,75,78,155
Land Rover 169
Large Blue 6,13,17,19,20,71, 99,107,134,140,141
Large Blue Action Plan 173
Large Blue Committee 72
Large Copper, 3,13,17,20,33, 40,55,132,134
Large Heath 33,143,147
Large Tortoiseshell 29,33,140, 159
Large White 4
Lasius ants 120
Leguminosae 5
Lhonoré, J. 9, 95
limestone grassland 100,119
linkage 100
Linnean Society 32
LNR 125,164
local reserves 166
Lodmoor 142
London 166
Lotus corniculatus 120
Lowe, David 42
Lugg meadows 119
Lulworth Cove 16
Lulworth Skipper 3, 33,100,159
Lycaena dispar batavus 56,62
Lycaena dispar bordielensis 62
Lycaena dispar carueli 56
Lycaena dispar gronieri 56
lycaenids 6

Maculinea arion arion 72
Maculinea arion eutyphron 91
Maculinea genus 88
Magdalen Hill Down 104
management plans 107
Map butterfly 159
Mar Lodge Estate 108
Marbled white 12,30,104,122,143
margins 105
Marsh Fritillaries 33,34,44,99,101 116-117,134-135,137,144-145
Martlesham Heath 121
Maryport Harbour SSSI
Mazarine Blue 13,40
McLean, I. 165
Meadow Brown 30,85,104,165
Meadows 118-119
metapopulations 86
migrants 27
Mike Brown reserve 166
Milk-parsley 46
Moberly, J.C. 46
MOD 15,37,106,115
Monks Wood 49,60,64.100,124,146
Moore, Norman 41,80
Mountain Ringlet 2
Mr X 41
Muggleton, John 84,88,89
Myrmica sabuleti 78,83
myxomatosis 76, 83-4, 86,105,114

National Lottery 108
National Nature Week 42
National Parks 118
nature reserves 163
NC 17
NCC 17,27,28,108,115
NCC contracts 21
Newman, L.H. 9,36,159
Newman, L.W. 12
NGO 32,34,164
NHM 72
NNRs 31,125,145,148,164
Northern Brown Argus 2,141
NRIC 15
NT 14,15,48,80.103,107.108, 118,142,155
NT Scotland 108
Nymphalidae 5

Oates, M. 161
Oberthür, Charles 58,96
Old Winchester Hill 105,147
Operation Butterfly 42
Operation Neptune 107
Orange Tip 148,165
Ornithoptera 131
Ouse Washes 132
Outline Planning 121

Painted Lady 3
Papilio machaon britannicus 46,140
parasites 61,64
parasitoids 88
Park Slip West 116
patches 85
Peachey, Caroline 29
Peacock 27,42,165
Pearl-bordered Fritillary 33,114
pesticides 124
Phryx vulgaris
Plantlife 32,125,126
Plebejus argus caernensis 140
Polak, R.A. 58
Pollard, Ernest 20,99
Populus tremula 58
Portland 155
predators 64
preservation 6-8
Project Papillon 42
Public Authorities 106
Pullin, Andrew 56,65
Purefoy, E.B. 57-59,68,87,90
Purple Emperor 30,99,102-104, 109,141
Pyle, R.M. 162

Queen of Spain fritillary 131

rabbits 104
Ramsar sites 131,134
Rangnow, H. 57

Ratcliffe, Derek 91
RDB 92,140
Red Admiral 27,42
Red Data Books 138
regional differences 99
requisitioning 15
RESL 16,73,78
richness 28
rides 103
Riley, N.D. 13,56,73,155
ringlet 141
Rio Convention 30
road impact 122
Roper, P. 162
Rothschild, Lord 11,12,55,58,63, 90,154
Rothschild, N. C. 11,12-13,14,29, 58,59
Rothschild, Miriam 15,27,29,55, 74,75,129
Rothschild, Victor 17
Rowell, T.A. 143
RSNC 12,32,48,106
RSNC introduction policy 178
RSPB 108,132,164,171
Rumex crispus 62
Rumex hydrolapathum 56
Russel, Freddie 75

SAC 136
Sainsbury plc 121
Salix caprea 103
Salix cinerea 103
Salix fragilis 103
Salix sp. 58
Schofield, J. 59
Scotch Argus 2-3,98,159
Scottish butterflies 2
SCPBI 72-3,78
Selar Farm Grassland 117
set-aside 133
Shakespeare Cliffs 122
Sheldon W.G. 13
Sheppard,. D. 62,146
Shetlands 40
Shoard, Marion 28
Short, H.G. 62
Short-tailed Blue 131
Silver-spotted skipper 33,101, 114,140
Silver-studded Blue 40,98,100, 101,141,120-121,144
Silver-washed Fritillary 29,102
Site M 93
Site X 87
Site Y 87
skippers 5,104,108
Small Blue 118,141
Small Copper 165
Small Heath 165
Small pearl-bordered Frit. 114

Small Skipper 30
Small Tortoiseshell 27,42,98,165
Small White 4
Scottish National Heritage 131
Southern England 106,114
Southwood, T.R.E. 43
SPA sites 134
Species richness 31
Speckled Wood 30,122,165
Spencer, Earl of 30
SPNR code 12,160
Spooner, 76,80,83
spraying 123
SPA. 18, 66,88,92
SSSI 29,31,32, 113,115-116, 125,142,143
St Malo 91
Stamp, Dudley 80
Stubbs, Alan 141,146
Suffolk Trust 120-121
swaling 83
Swallowtail, scarce 2, 9
Swallowtail 3,13,20,33,39, 45,99,132,141,155
Swedish blues 90,92-94
Sylvia communis 64

Tansley, Professor 25
Thomas, Jeremy 41,72,75,100
Thomas, C.D. 97,101
threatened species 108
Threats 4,113
Titchwell 133
Tony Steele reserve 166
Tor grass 101
Turloughs 151
Tutt, James 56,57

UK laws 140
Ulex europeaus 120
Urticaceae 5

Vane-Wright, R. 86
Verity, Roger 91
Verrall, G.H. 57
Violaceae 5

Wall 42,165
Warren, Martin 19,52,97,99, 101,106
Warren, M. 114,116,161
WCA 1981 32,44,92,115,141
Webb, Mark 56,65
Whalley, Paul 34
White Admiral 99
White-letter hairstreak 122
Whitethroat 64
Whixall Moss 143
why conserve? 27
Wicken Fen 46,47,51,56, 87, 108,133

Wildlife Link 12,16
Wildlife Trust's reserves, The 192
Wildlife Trusts, The 12,164,166
Williams, Mike 143
Willmott, Ken 99,102
Wittpen, J.H.E. 59
Wood White 13,99,114,141
woodland edge 99,103
Woodland Trust 106
woodlands 36
Worcestershire 119-120
World Heritage Convention 130
WWF 32,92,102,108

Yorkshire 105,114,166